Man Nature and Environment

(NEHU Syllabus)

(For B.A., B.Sc. & B. Com. Students)

Anil Kumar De
Formerly, Professor and Head of Deptt. of Chemistry
and Dean of Science Faculty, Visva-Bharati University
West Bengal

Arnab Kumar De
Senior Lecturer in Chemistry
Raja Peary Mohan College, Uttarpara
West Bengal

NEW AGE INTERNATIONAL (P) LIMITED, PUBLISHERS
New Delhi • Bangalore • Chennai • Guwahati • Hyderabad
Kolkata • Lucknow • Mumbai

NEW AGE INTERNATIONAL (P) LIMITED, PUBLISHERS
4835/24, Ansari Road, Daryaganj,
New Delhi - 110 002

Offices at:
Bangalore, Chennai, Guwahati, Hyderabad, Kolkata,
Lucknow and Mumbai

ISBN : 81-224-1452-4

Published by K.K. Gupta for New Age International (P) Ltd.,
4835/24, Ansari Road, Daryaganj, New Delhi-110 002 and
printed in India at De Unique, New Delhi-110 018.

Preface

The General Foundation Course (GFC) on Environment was introduced by the North-Eastern Hill University (NEHU) in 1983 at the undergraduate level. The Honourable Supreme Court in their landmark judgement in 1995 issued a directive to the Union Government for introducing environmental education at the undergraduate levels for arts, science and commerce streams. The University Grants Commission, in turn, urged all the universities to introduce environmental course at undergraduate level as a common course for students of arts, science and commerce streams. NEHU was ahead in introducing a course on Man and Environment from 1998 onwards while other universities followed. However, there was confusion all around due to lack of appropriate textbooks and trained faculty with the result that the students did not take interest in the subject. In this scenario the present book on Man, Nature and Environment has been designed to meet the challenging requirements of a common course and fill the urgent need of the academic community.

The present book covers the NEHU syllabus for generating awareness among the students about environment and environmental issues in general and regional environment, issues and remedies, in particular. It is organised into ten units as per syllabus in a logical sequence. Keeping in view the diverse backgrounds of the students (Arts/Science/Commerce), science has been kept at a low key and the subject matter is presented in a highly lucid and attractive style so as to inspire the young minds. The Glossary in Appendix 1 is intended to reinforce the understanding of the students of all streams about the key terms and concepts in Units 1-10. Unit 11 presents Feedback Exercise (Objective tests) which will be found useful by the students and teachers alike. An additional section on NEHU Questions 2000 (3rd year exam) will also be important reference for them.

The authors wish to record their grateful thanks to Mr. Rajiv Banerji, General Manager (Publishing) and Mr. K.K. Gupta, Managing Director, New Age International (P) Ltd. Publishers, New Delhi for their kind offer and prompt publication of the book. Constructive suggestions and comments for further improvement of the book in its next edition are cordially invited from fellow teachers and students of the N.E. region. Our sincerest thanks are due to Mrs. Kalpana De, Chandrima and Charchita for their sustained encouragement.

Anil Kumar De
Arnab Kumar De

Contents

Unit 1

Basic Concepts

1.1. WHAT IS ENVIRONMENT?

There is much concern about environment nowadays all over the world. It started in the developed countries in 1960s and swept across India in 1980s. The Bhopal Disaster of 1984 sounded the wake-up call—the worst industrial disaster in history killed 10,000 people in one week and sent shock waves throughout India. It focussed on the need of environmental consciousness at all levels.

First of all, we have to understand what is meant by environment. According to some, environment means the surroundings around us. *Environment* is broadly defined as the sum total of all conditions and influences that affect the development and life of all organisms on earth. The living organisms vary from the lowest micro-organisms such as bacteria, fungus, etc. to the highest including man. Each organism has its own environment.

The *functions* of the environment in relation to man are: (a) recreation, (b) natural resources — agricultural, mineral and forestry which man consumes directly or indirectly; (c) "sink" for receiving wastes produced by man and his activities. These functions of the environment are damaged by human activities which subject the environment to four stresses: (i) "eutrophic" i.e. the task of decomposing wastes produced by consumption and production activities; (ii) "exploitative" i.e. cropping of plants, extraction of minerals and hunting of animals; (iii) disruptive — brought about by activities such as deforestation, construction of highways and towns and (iv) "industrial" stress which results from industrial development.

1.2. ELEMENTS OF ENVIRONMENT (ENVIRONMENTAL SEGMENTS)

The environment consists broadly of two components: Non-living or Physical and Living or Biological.

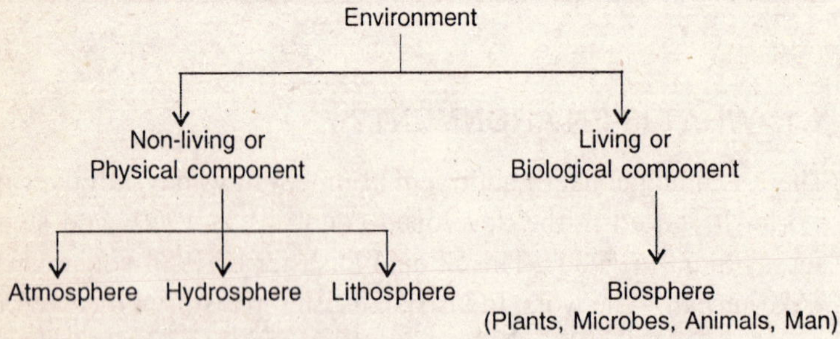

(1) **Atmosphere:** The atmosphere has three major types of constituents —major, minor and trace. Pure (pollution-free) dry air at ground level has the following components, expressed in percentage by volume (within brackets):

Major components	:	Nitrogen (78.09)
		Oxygen (20.94)
		Water vapour (0.1)
Minor components	:	Argon (0.9)
		Carbon dioxide (0.032)
Trace components	:	Neon (0.0018)
		Helium (0.0005)
		Methane (0.0002) etc.

The properties (parameters) of the atmosphere vary much with altitude. The density shows sharp decrease with increasing altitude. Pressure drops from 1 atmosphere at sea level to 3×10^{-7} atmosphere at 100 km above sea level while temperature varies from –92° to 1200°C. The total mass of the atmosphere is about

5×10^7 tons which is roughly one millionth of the earth's total mass 5×10^{24} kg.

Structure: The atmosphere may be broadly divided into four regions as shown in the Table 1.1 below. It extends up to 500 km with temperature varying from a minimum of –2°C to a maximum of 1200°C.

Table 1.1: Major regions of the atmosphere

Region	Altitude range km	Temperature range °C	Important chemical species
Troposphere	0-11	15 to –56	Nitrogen Oxygen Carbon dioxide Water vapour
Stratosphere	11-50	–56 to –2	Ozone
Mesosphere	50-85	–2 to –92	$Oxygen^+$ Nitric oxide
Thermosphere	85-500	–92 to 1200	$Oxygen^+$ Nitric $Oxide^+$

$Oxygen^+$ = Oxygen atom with positive charge

Nitric $oxide^+$ = Nitric oxide molecule with positive charge. These molecules/atoms pick up charge in the upper atmosphere

The *Troposphere* contains 70 per cent of the mass of the atmosphere. Density decreases exponentially with increasing altitude. The temperature decreases uniformly with increasing altitude (negative lapse rate). In this region the air masses are constantly in circulation as energy flows due to imbalances in heating and cooling rates between equator and the poles.

In contrast to the Troposphere, the *Stratosphere* is the quiet layer having a positive lapse rate i.e. increase in temperature with increase in altitude. Here ozone is the important species which absorbs ultra-violet radiation and raises the temperature. The ozone layer serves as a protective shield for life forms on earth from the harmful effects of the sun's ultra-violet radiation. Because of slow mixing in the stratosphere, the molecules or particles in the region stay for a long time. If some pollutants enter this layer, they will stay for a long period and slowly come down to the troposphere causing long-term global hazards. Introduction of nitrogen oxides by jet planes and refrigerant gases (chlorofluorocarbons) leads to thinning of ozone layer and generates ozone hole.

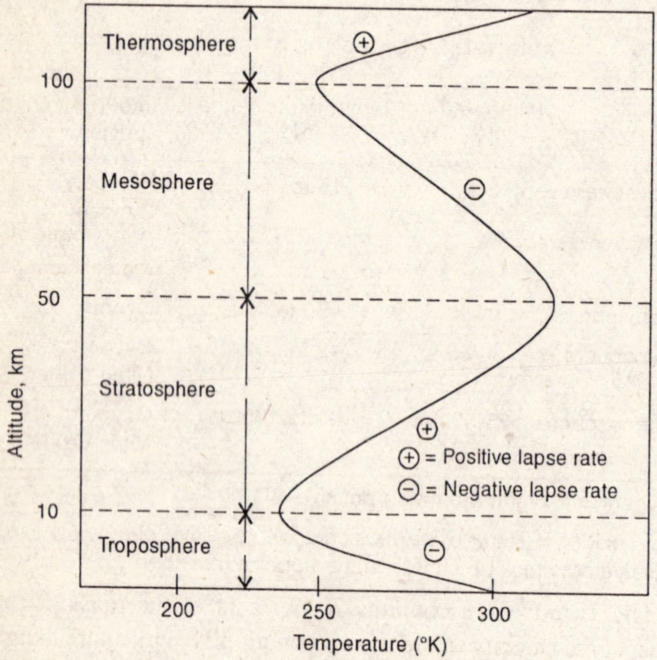

Fig. 1.1. Major regions of the atmosphere

1°K = −273°C;

Negative lapse rate means temperature decreases with increase in altitude
Positive lapse rate means temperature rises with increase in altitude

In the *mesosphere* temperature decreases with increase in altitude. This is due to low levels of ultra-violet species i.e. ozone.

In the *thermosphere* temperature rises once again giving a positive lapse rate. Here oxygen and nitric oxide ionize after absorption of solar radiation in the far ultra-violet region.

(2) **Hydrosphere**: The history of human civilisation shows that water supply and civilisation are intimately linked with each

other. Several cities and civilisations have vanished due to shortage of water supply.

The world's total quantum of water is 1.4 billion km^3. If all the seabeds could be filled up and brought at the level of the earth's surface, then the entire water in the seas would cover the earth's surface and make it 2.5 km-deep water mass. About 97 per cent of earth's water supply is in the ocean which is unfit for human consumption and other uses due to high salt content. Of the remaining 3 per cent, 2.3 per cent is locked in the polar ice caps and hence out of bounds. The balance 0.7 per cent is available as fresh water but the bulk of it is ground water, 0.66 per cent and the rest 0.03 per cent is fresh water in rivers, lakes and streams. The break-up of this 0.03 per cent fresh water is: lakes and ponds 0.01 per cent, water vapour 0.001 per cent, rivers 0.003 per cent and water confined in plants, animals and chemicals 0.0187 per cent. [United National Water Conference Report, Argentina, 1977]

Thus we see that we have a very limited stock of usable water, 0.03 per cent surface water (rivers, streams and ponds) and 0.66 per cent ground water. The quantity of water vapour arising from evaporation of sea water and river water returns by the same volume to the earth's surface by rainfall and back to the water sources. This natural hydrological cycle is more or less balanced in terms of charge (cloud formation) and discharge (rainfall) and is the source of fresh water supply. But we are drawing large quantities of ground water for agriculture and industries while the waste water from these is much polluted and on mixing with rivers is polluting the rivers also.

Water is essential for all life forms on earth — plants, animals and man. The Hydrological cycle helps in exchange of water between air, land, sea, living plants and animals. It is based on massive evaporation of water from the seas and oceans, cloud formation and condensation into rainfall. It ensures continuous circulation of water between the oceans, atmosphere and biosphere (plants, animals and man). Thus we get our supply and reserves of fresh water. Fig. 1.2 illustrates the hydrological cycle.

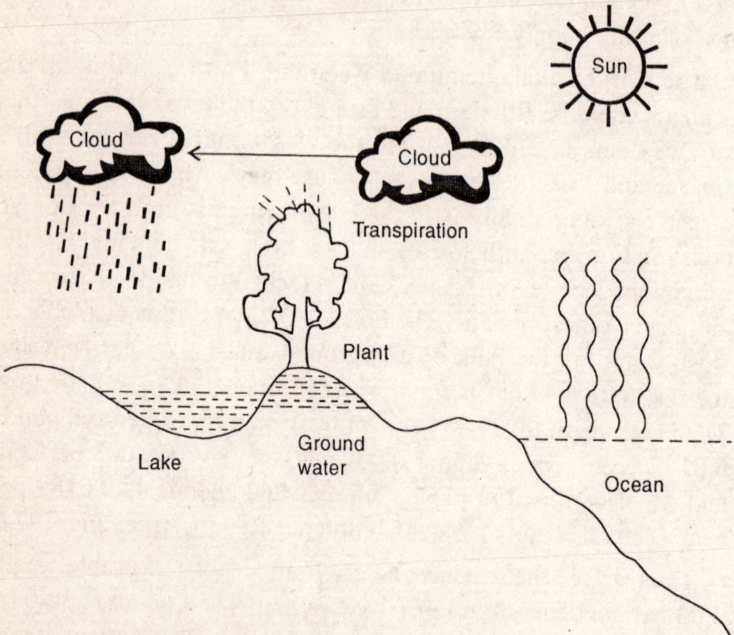

Fig. 1.2. The Hydrological Cycle

(3) **Lithosphere:** It is made of the mantle of rocks. It includes the soil which covers the rock's crust in many places. Rocks are subjected to continuous weathering forces—rain, wind, chemical and biological and suffer disintegration. The resulting primitive soil is suitable for the growth of plants—after death and decay, plant debris returns to soil. The mineral component of soil comes from the parent rocks by weathering processes while the organic component is due to plant biomass as well as populations of bacteria, fungi and insects (earthworms). A typical soil, suitable for agriculture, contains about 5 per cent organic matter and 95 per cent inorganic matter. Soil has an important role as it produces food for man and animals. Good soil and good agriculture are valuable assets for a nation.

In general, soil has a loose structure consisting of solid mineral, organic matter and air space. It shows broadly three zones as its depth increases. The top layer, up to several inches thick, is known as top soil which is an index of soil quality. This is the layer of maximum biological productivity and it contains bulk of organic matter. Hence it is very important for vegetation cover and agricultural crops. Reckless deforestation causes loss of top soil which also means loss of agricultural production. The underlying layer is the sub-soil which receives organic matter, salts and clay particles leached from the top soil. The third layer (zone) consists of weathered parent rocks from which soil was formed. Plants draw water and nutrients from soil—they transport water into the plant body—(roots and leaves) and then excess water into the atmosphere through leaves by the process of transpiration.

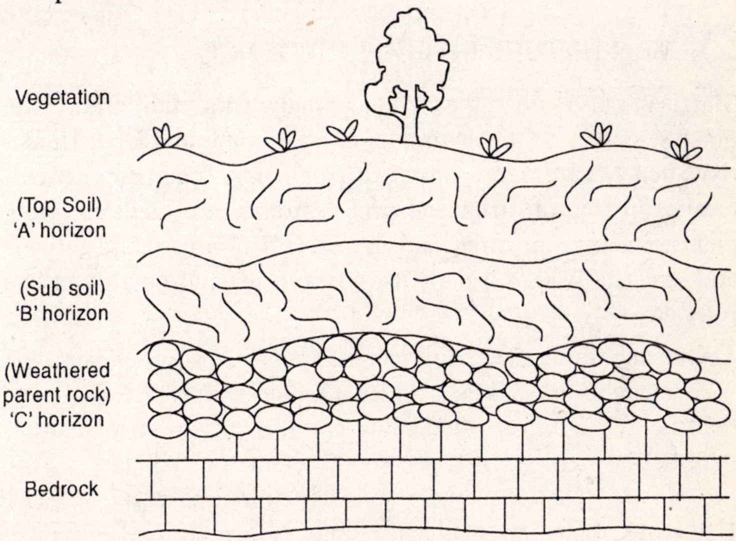

Fig. 1.3. Soil profile showing soil sections

The various land forms of the lithosphere are (i) Mountain, (ii) Plateau and (iii) Plain. Mountains provide natural frontier, shelter for tribals, important flora and fauna. They contain forests which are important natural resources. Plateaus are rich in forest and mineral wealth while the plains account for maximum world population because of convenience for cultivation, communication, transport and industrial growth.

(4) **Biosphere:** Broadly speaking, the biosphere consists of the earth's crust, hydrosphere, atmosphere and various living species (micro-organisms to man) which exist in the zone 600 metres above earth's surface and 10,000 metres below sea level. Both biosphere and environment have close interactions with each other. Thus oxygen and carbon dioxide levels of atmosphere depend entirely on the plant world. Green plants are responsible for accumulation of oxygen in the atmosphere through photosynthesis and decay. In primitive stage atmosphere was devoid of oxygen and there was no life form on earth. In general, the biosphere is closely related to energy flows in the environment and water chemistry.

1.3. MAN-NATURE INTERDEPENDENCE

Man is at the centre of the biosphere in dynamic equilibrium with other segments of the environment—air, water and land. He derives benefits from the natural environment—food from agricultural crops in soil, fruit and timber from trees, medicine from plants, meat from birds and animals, fish from ponds, rivers and seas and water for daily use from springs, ground water and river etc.

Our ancestors used to worship nature—Sun, water, air, soil, various plants and animals. Ancient Indian philosophers believed that the entire universe including earth, plants, animals and human beings is made of five essential elements viz. *khiti* (earth), *op* (water), *teja* (fire, energy), *marut* (air) and *byom* (space). Bodies of plants, animals and human beings are created out of these five elements and after their death and decay, their bodies are disintegrated and converted back to these elements.

The fundamentals of Conservation Ethics were brilliantly formulated in the *Ishopanishad* during the Vedic era:

> "The whole universe together with all its creatures belong to the Lord (Nature). Let no one species encroach over the rights and privileges of other species. One can enjoy the bounties of Nature by giving up greed." In other words, if Nature is overexploited we shall lose all the benefits.

In the Maurya era the emperors emphasised the awareness campaign for conservation of Nature and non-violence towards trees and animals. India has a rich tradition of environmental ethics as recorded in our ancient scriptures and holds the model for the present age.

Primitive man lived in mountain caves and looked at the forces of Nature with wonder and awe. He was panicky about the Sun, stars, lightning, thunder, rain and also wild animals in the forests. He had no tools to overcome the hazards from Nature. Gradually he invented fire and stone tools for his defence against wild animals. Agriculture was discovered some 10,000 years ago and that was the starting point for the march of human civilisation. Man devised science and technology step by step and depended more and more on these for improving his living standards. This became prominent after the Industrial Revolution (1760-1840) which transformed England from an agricultural country into a predominantly industrial one. The invention of steam engine by James Watt (1785), mechanisation in textile industry, advancement in mining, transport and ship-building industries made Britain the most prosperous country in the world. This was promoted by British colonial exploitation. While man possessed the technology to establish his supremacy over Nature, the environment became more and more polluted and degraded. Man-made activities led to rapid deterioration of the quality of environment and at the same time it threatened the survival of man himself on earth.

1.4. ECOSYSTEM AND ECOLOGY

The word "*Ecology*" was coined by a German biologist in 1969 and was derived from the Greek word "*Oikos*" meaning "*House*". Ecology is the branch of science that deals with the study of interactions between living organisms and their physical environment. The latter are closely inter-related and they have continuous interactions so that any change in the environment has an effect on the living organisms and vice-versa. Any unit or biosystem that includes all the organisms which function together (biotic community) in a given area where they interact with the physical environment is known as *ecosystem.*

The ecosystem is the functional unit in ecology as it consists of both the biotic community (living organisms) and abiotic environment. The latter have close interaction, essential for maintenance of life processes. The interaction is conducted by energy flow (solar energy) in the system and cycling of materials (natural cycles—hydrological cycle, nitrogen cycle, oxygen cycle, phosphate cycle etc.)

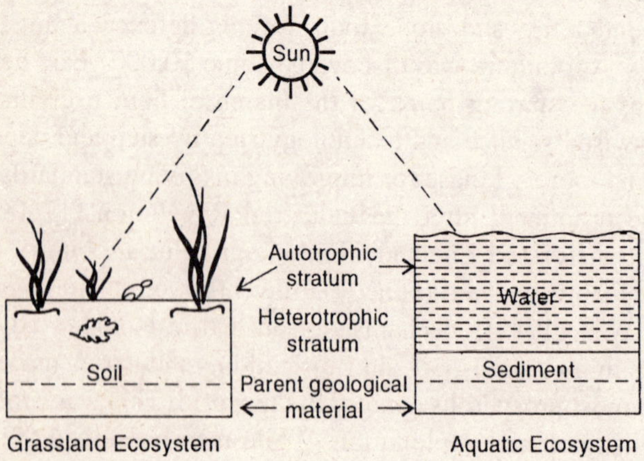

Fig. 1.4. Grassland and Pond Ecosystem

From biological point of view, the ecosystem has the following constituents:

(i) inorganic substances (carbon, nitrogen, carbon dioxide, water etc.) involved in natural cycles

(ii) organic compounds (proteins, carbohydrates, humic substances etc.)

(iii) air, water and substrate environment including the climatic regime and other physical factors

(iv) producers, autotrophic (self-sustaining organisms) green plants that can manufacture food from simple inorganic substances

(v) heterotrophic (i.e. depending on others for nourishment) organisms or particulate organic matter

(vi) micro-consumers, decomposers mainly bacteria, which obtain their energy by breaking down dead tissues or by absorbing dissolved organic matter, extracted from plants or other organisms. The decomposers release inorganic nutrients that are utilised by producers. They also supply food for macro-consumers or heterotropic organisms (bacteria, fungi and animals) and often excrete hormone-like substances that inhibit or stimulate other biotic components of the ecosystem.

The common features of all ecosystems—terrestrial, freshwater, marine and agricultural—are the interactions of the autotrophic and heterotrophic components. The major autotrophic metabolism occurs in the upper "green belt" stratum where solar energy is available while the intense heterotrophic metabolism occurs in the lower "brown belt" where organic matter accumulates in soils and sediments.

QUESTIONS

1. What is meant by Environment?
2. What are the functions of Environment in relation to Man?
3. Name and describe the Environmental Segments.
4. What are the major and minor components of atmosphere?
5. Give an account of water resources of the world giving their percentages.
6. Illustrate the Hydrological Cycle.
7. Show the soil profile with the various zones.
8. What is Ecology?
9. Explain the term Ecosystem with an example.

N.B. For Objective Tests see Unit 11.

Unit **2**

Inter-Related Nature of Life on Earth

All living organisms in the ecosystem are made of matter and energy. Matter forms the building blocks of the physical and biological environments while energy provides the force to hold the structures together and helps the organisms to live. Energy is governed by the Laws of Thermodynamics. According to the First Law of Thermodynamics, the energy can neither be created nor destroyed but can be transformed from one form to another. The Second Law of Thermodynamics states that with each successive energy transfer or transformation, less energy is available to do work. It may be noted that in the process of transfer of energy from one organism to another, a large part of energy is degraded into heat and dissipated.

2.1. ENERGY SOURCE

The major input of energy on the earth's surface is from the Sun. Other sources of energy are from man-made activities e.g. geothermal energy, wind energy, tidal energy, nuclear energy and energy from fossil fuels. For all the living organisms in the biosphere the solar radiation is the main source of energy. The latter is transformed into food energy (a form of high quality chemical energy) by green plants bearing chlorophyll, algae and some bacteria through the process of *photosynthesis*. The energy passes from plants to animals when they eat green plants and the process goes on from one organism to another in the ecosystem. This is known as *Chemosynthesis*. However, this type of energy fixation through chemosynthesis amounts to a tiny fraction of the total energy flow in the ecosystem.

The Sun is a giant thermonuclear fusion reactor, which produces energy from fusion reactions of helium nuclei at high temperature and pressure, giving tremendous quantities of heat and light. The Sun radiates energy in the form of heat and light in all directions. Only a small part of this energy is intercepted by our planet Earth, located 150 million km from the Sun. The earth receives only $1/50 \times 10^6$ (one 50-millionth) of the Sun's energy output. Outside the Earth's atmosphere, the average intensity of solar energy is 1.2 kilowatt per sq. metre (1.2 kw/m^2). When this solar radiation passes through the atmosphere, a portion of it is absorbed by the atmospheric gases and particles. The net amount of solar energy reaching the Earth's surface is 0.9 kw/m^2.

The radiations reaching the earth's upper atmosphere are (wavelengths in nanometers shown within brackets):

(i) ultra-violet radiation (less than 400 nm), (ii) visible radiation (400-750 nm) and infra-red radiation (750-2500 nm). Among these ultra-violet radiation is harmful to the biosphere — flora, fauna and man. But the ozone layer in the stratosphere absorbs the bulk of the ultra-violet radiation so that only a very small part of it reaches the Earth's surface.

The major part of the solar radiation is occupied by visible light and heat wave, both of which contribute to our energy system. The solar radiation is mostly absorbed by land and water or reflected into space by water, snow and land surfaces.

A very small fraction of the radiation in the range of visible light is captured by organisms through photosynthesis. Further, half or more of the light absorbed by the leaves of plants is consumed for evaporating leaf water (*transpiration*). Consequently only about 1-2 per cent of the radiation that falls on the leaf surface is captured for photosynthesis. This small percentage of energy, in fact, represents the energy base for almost all life forms in the biosphere.

2.2. ENERGY PATH

The green plants transform the absorbed solar radiation into potential or food energy through the process of photosynthesis.

$$\text{Carbon dioxide} + \text{water} \xrightarrow[\text{chlorophyll}]{\text{sunlight}} \text{carbohydrate} + \text{carbon dioxide} + \text{water}$$

$$6\,CO_2 + 12\,H_2O \xrightarrow[\text{chlorophyll}]{\text{sunlight}} C_6H_{12}O_6 + 6CO_2 + 6\,H_2O$$

The rest of the biological world other than chlorophyll bearing plants and chemosynthetic bacteria derives its potential chemical energy from organic matters produced by either plant photosynthesis or chemosynthesis. An animal takes in chemical potential energy of food from the plant source, most of which is converted into heat to enable a small part of the energy to be re-established as the chemical potential energy of new protoplasm. During the transfer of energy from one organism to the other, a large portion (about 80-90 per cent) is degraded into heat at each step of the transfer. The transfer and dispersal of energy is guided by the *Stability Principle* which in turn follows from the Second Law of Thermodynamics. According to the *Stability Principle*, any natural enclosed system with energy flowing through it tends to change until a stable adjustment is developed through self-regulating or homeostasis mechanism. While matter in natural cycles (nitrogen cycle, oxygen cycle, phosphorus cycle etc.) moves in a cyclic path, energy moves in a unidirectional way in the ecosystem and then passes into the biosphere.

2.3. ENERGY FLOW/ENERGY BUDGET

The movement of energy through ecosystem is called *energy flow*. It is an essential part of the ecological process of life support in nature. As stated earlier, green plants transform solar energy with the help of carbon dioxide and water into food energy (chemical energy) by photosynthesis. This food energy goes on transforming when food is taken as energy by other organisms of different trophic levels. Thus energy flow is related with Food Chain. Energy, almost all from sunlight, is trapped by autotrophic organisms of the first trophic level. Because much energy is dissipated, about 90 per cent of available chemical energy is lost at each stage of transfer from one trophic level to the next higher level.

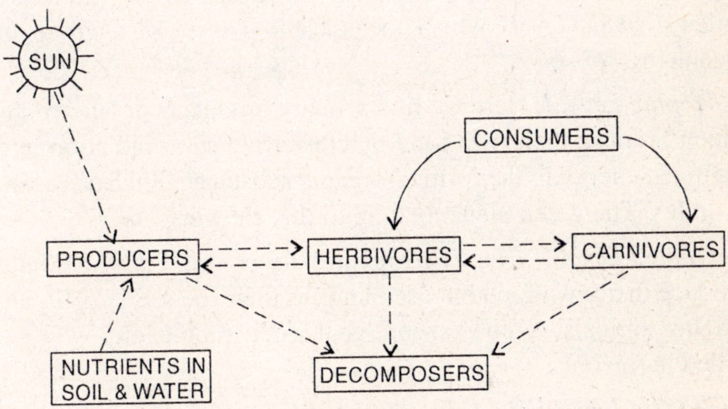

Fig. 2.1. Cycle of Energy and Nutrients Flow in Ecosystems

--> Direction of flow of energy — Non-cyclical

⟶ Direction of nutrients' flow — cyclical

Energy Budget

Energy Budget shows a record of the flow of energy through an ecosystem in each step from one organism to another.

The *primary productivity* of an ecosystem is defined as the rate at which radiant energy is converted by photosynthetic and chemosynthetic activity or producer organisms (green plants, for example) to organic substances.

In coastal waters of seas primary production in relation to biomass shows vertical distribution. It is concentrated in the upper 30 meters or so — in the clearer but less fertile waters of the open sea the primary production zone may extend down to 100 metres or more. This is why coastal water appears dark greenish and the ocean waters blue.

2.4. TROPHIC LEVELS

The transfer of food energy takes place in some hierarchical order in the ecosystem. The food energy passes from one group of organisms to some groups at different levels. These levels are called *Trophic levels* which mean feeding level of a group of organisms.

Trophic Level I: Here we find primary producers or autotrophs which produce their own food by converting solar radiation into chemical energy in the form of organic substances such as carbohydrates. The green plants belong to this category.

Trophic Level II: Here we find primary consumers who do not produce their own food but use plants as their food. Examples are grazing animals — cow, sheep, goat, deer, rabbits etc. who are called herbivores.

Trophic Level III: At this level we find animals who depend upon other animals for their food. They are carnivores (tiger, lion, etc.) who eat mainly herbivores. This is the secondary consumer level.

Trophic Level IV: It is the tertiary trophic level to which belong all groups of animals including man who obtain their food from the other three levels. They are omnivores who eat both plants and animals.

It may be pointed out that organisms deriving their food energy from the same producer level belong to the same trophic level. Thus grasshopper, cattle and grain-eating birds are all primary consumers and occur at the second Trophic level since they derive their energy from plants.

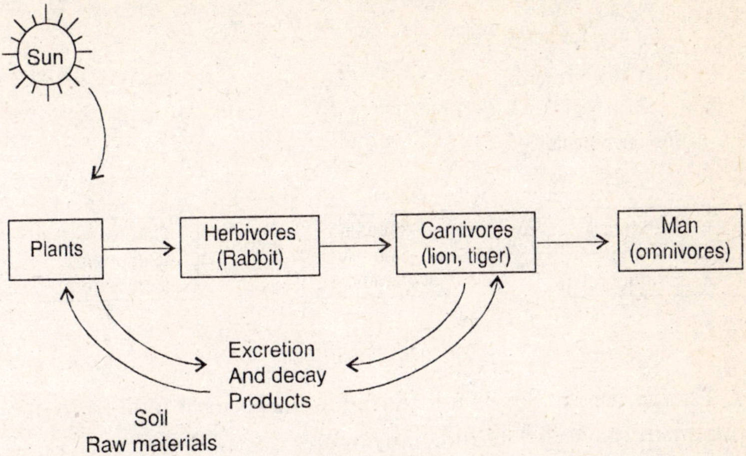

Fig. 2.2. Relationship among different biotic components of an ecosystem

2.5. FOOD CHAIN AND FOOD WEB

Food chain means the transfer of food energy from the source in plants through a series of organisms with the repeated process of eating and being eaten. In a grassland, grass is eaten by rabbits while rabbits are in turn eaten by foxes. This is an example of a simple food chain. This food chain implies the sequence in which food energy in transferred from producer level to consumer or higher trophic level. It has been observed that at each level of transfer, a large proportion, 80-90 per cent, of the potential energy is lost as heat. Hence the number of steps or links in a sequence is restricted, usually to four or five. The shorter the food chain or the nearer is the organism to the beginning of the chain, the available energy intake is greater.

Basically, food chains are of two types: (i) *Grazing food chain* and (ii) *Detritus food chain*. The grazing food chain starts from green plants and passes through herbivores to carnivores. On the other hand, the detritus food chain starts from dead organisms and goes into the micro-organisms and then to detritus feeding organisms and their predators. The term '*detritus*' refeıs to all organic matter involved in the decomposition of dead organisms.

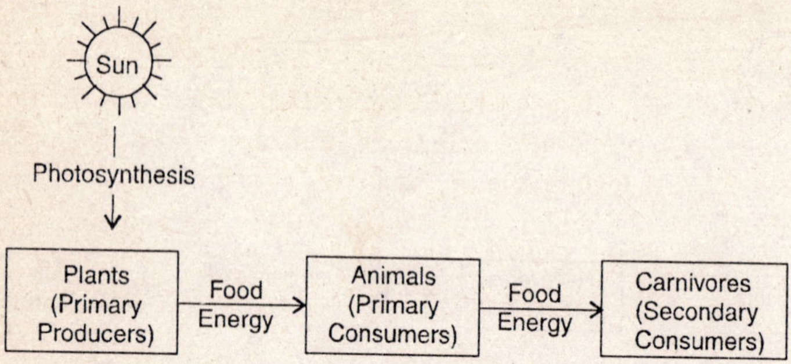

Fig. 2.3. Food Chain

Energy for the Sun is transferred in the form of food from one organism to another in the ecosystem.

Food Web

In complex natural communities the food chains are not found in isolation. They are all inter-connected with each other since most consumers have multiple food sources and many species are prey to several predators and parasites. Such inter-locking food chain is called *food web*.

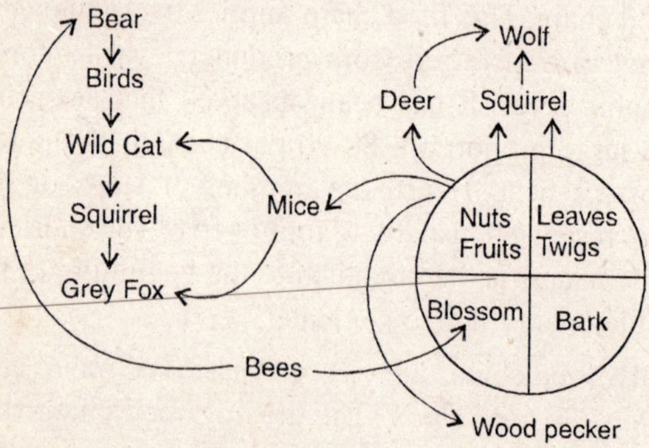

Fig. 2.4. A Food Web in a terrestrial ecosystem

2.6. CONCEPT OF COMMUNITY

In an ecosystem different kinds of organisms live and grow in association with each other. Plants, animals and microbes living together with mutual adjustment and interdependence in a prescribed area or habitat form a community. In other words, communities may be defined as groups of interacting populations. Thus when only plant populations are considered, the community is called plant community. Animal community means assemblage of animal populations. In a community there may be hundreds or thousands of kinds of organisms but they share the same habitat growing in a uniform environment. Examples of natural communities are a forest, grassland, desert or a pond.

A community has several important characteristics as follows:

(1) *Species diversity* — A large number of species are rare but important and they determine the species diversity of the community.

(2) *Dominance* — A few species are the most important ones in respect of the nature, energy flow, and stability of the system. They are said to be the dominant species.

(3) *Growth form and structure* — Within a community several growth forms such as herbs, shrubs, trees etc. can be identified. besides canopy architecture and shape are also important determinants of growth form.

(4) *Stratification* — It refers to the vertical and horizontal layering of organisms in a community. In a forest community vertical stratification may be distinguished as herb, shrub, subcanopy, canopy and emergent layers.

(5) *Succession* — The biotic communities are constantly undergoing replacement of one species by another along a time gradient in absence of any external interference.

Population and community are two distinct ecological terms. By *population* is meant a collective group of organisms of the same species in a particular habitat. Examples are human population, tree population, fish population etc. On the other hand, a *community* consists of many kinds of species and is known by the name of their physical habitat such as forest community, pond community etc. A community is, in general, a self-sustained unit of producers, consumers and decomposers.

QUESTIONS

1. Distinguish between Photosynthesis and Chemosynthesis.
2. How do green plants produce their own food?
3. How does the flow of energy in an ecosystem conform to the Law of Thermodynamics?
4. How does food energy pass from one group to another group of organisms?
5. Show how solar energy is transferred from one organism to another.
6. Illustrate Food Web in an ecosystem.
7. What do you mean by Community? How do you distinguish it from population?

For Objective Tests *see Unit 11*.

Unit 3

Population and Environment

3.1. HUMAN POPULATION AND DISTRIBUTION

Population is intimately related to environment. The Human population has grown faster in the 20th century than ever before. World population doubled in 40 years between 1950 and 1990 to cross five billion. The developed countries account for 1.5 billion and developing countries 3.5 billion population. By 2000 AD the population has touched 6.3 billion and by 2010 it will grow to 7 billion (one in every seven will be an Indian). World population is growing by 92 million every year, roughly adding population of Mexico.

It is interesting to note that it took about 2 million years for the world population to become 1 billion (1830), 100 years for 2 billion (1930), 30 years for 3 billion (1960), 15 years for 4 billion (1986) and 11 years for 5 billion (1997). The population stands at 6.3 billion (2000) and is estimated to be about four times around 22.5 billion in 2100. In developed countries the population is likely to be less than double while in developing countries like India about four times (2100).

The statistics for India is of serious concern. Between 1901 and 1951 India's population grew from 238 million to 361 million an increase of 52 per cent in 50 years. Between 1951 and 1981 it expanded from 361 to 685 million. Post-independent India in 35 years (1947-81) literally added a second India, i.e. doubled its population. At present it has touched 1 billion mark and is the second most populous country, next to China (May 11, 2000). We have been overwhelmed by population explosion since 1980.

Distribution

For historical and other reasons people are not uniformly distributed. USA and Canada have a population of 250 million; South America and the Soviet Union (CIS) have the same population. Africa and Western Europe have about 500 million people; East Asia, i.e. China, Japan and Korea have more than 1 billion while South Asia is the most populous region, 1.5 billion (India, Bangladesh, Pakistan more than 1 billion). India is adding every year the population of Australia at the current rate.

Population Density

The density of population within a country gives a rough idea of the effect of population on natural resources. In general, lightly populated countries retain more of their original vegetation and wild animal populations than those with dense human population. Thus Australia (4 people/sq. km) is not likely to face wood or water shortage compared to Bangladesh (1800 people/sq. km). However, total population size or population density cannot accurately predict the impact of environmental damage on the economy of a country. Some of the most crowded countries in East and South Asia are the most prosperous, e.g. Taiwan, South Korea, Hong Kong and Singapore where the population density is 1000 people/sq. km.

Some crowded countries can even manage to produce all their own food. U.K. (600 people/sq. km) exports food while Brazil (44 people/sq. km) has to import food. Again, Europe has all the timber and fresh water it requires but Africa (95 people/sq. km) has acute shortage of both these items.

Some differences among countries as stated above are due to differences in their economies—sustainable and non-sustainable. The countries with stable population enjoy sustainable economy and are prosperous—the developed countries belong to this category. Similarly countries with fast population growth suffer from non-sustainable economy and face serious problems.

Age Structure

In contrast to the developed countries India has a pyramidal distribution of age-wise population distribution. Children constitute 40 per cent, youth 33 per cent, middle-aged persons 21.5 per cent and old 6.2 per cent. Such a structure with predominantly young people up to 34 years (73 per cent) favours fast population growth. While India has a youthful age (reproductive group) structure, in the Western countries the population has much less youthful age structure which slows the population growth.

The ratio of people over 65 and under 15 to the rest of the population is *dependency ratio*. This gives a measure of the economic impact of the population age structure. People above 65 and below 15 contribute little to the economy and must be supported by the working population. Dependency ratios are rising in developing countries. High dependency ratios, as in India, have adverse effect on the economy. This explains why working Indians have lower standard of living compared to developed countries.

Doubling Time

The period within which the population of a country doubles is known as its doubling time. This gives some idea of the adverse effects of population growth. Most developed countries have doubling times of more than 100 years while in developing countries, as in India, population has doubling time less than 25 years. It is absurd to double our resources—water, food and energy in 25 years and tackle the pressure on schools, hospitals, police protection and other vital services.

Fertility Rate

The general fertility rate of a population is the number of babies born to 1000 women of the reproductive age. The rate of population growth, however, is based on the average age of first reproduction. Thus whether a population is increasing, decreasing or stationary is more correctly estimated from the age-specific fertility rate, the number of births per 1000 women of each reproductive age group. The fertility rate for India and other developing countries is about 3.0 while that for the Western countries is below 1.5.

Infant Mortality/Life Expectancy

The death rate is generally expressed as the number of deaths per year for every 100 people. Modern medical science has helped to reduce the death rate but not the birth rate. Death control without birth control has led to population explosion.

Infant mortality is one of the most tragic indicators of poverty of a country. It also sets the parameter of *life expectancy*, i.e. the average number of years that a new-born baby is expected to survive. Progress in medical science has in general increased the life expectancy. In 1900 life expectancy was 46 years in the developed countries because the infant mortality was high, i.e. 40 per 1000. The situation improved much in 1984 in these countries when the life expectancy rose to more than 70 years as the infant mortality dropped to about 10 per 1000. The world's highest life expectancy is that of Japan (79 years) which enjoys highest literacy and fairly high standard of living. In the developing countries high infant mortality accounts for lower life expectancy. Furthermore, lack of the basic amenities of life (access to clean water, nutritious food and elementary education) is responsible for life expectancy of 45 years in most of Africa whereas in South America and East Africa life expectancy rose to 65 years.

It is on record that in developing countries 1 in 3 children is malnourished and about 3 million children die annually from diseases that could be avoided by immunization. And moreover, one million women die each year from preventable reproductive health problems.

Carrying Capacity

The maximum population size that an eco-system can support under particular environmental conditions is known as the *carrying capacity*. In natural ecosystems with unlimited resources and ideal environment, species can multiply at a maximum rate. However, in actual practice, the population of a species remains in check due to interaction of the inhabiting species as also finite nature of resources.

It is an established fact that while foodgrain production can under optimum conditions increase arithmetically (1, 2, 3, 4, ...), population increases geometrically (2^2, 3^2, 4^2, ...).

$$\frac{dN}{dt} = rN$$

where dN/dt = rate of population growth

N = population size, and

r = specific growth rate

With doubling of population, resources do not double and hence set in a critical situation. If the population size far exceeds the carrying capacity by a wide margin, it leads to *population crash* when environmental conditions get degraded and lower the carrying capacity.

The earth weighs about 5.97×10^{24} kg. If the present population growth continues for the next 5000 years, then the weight of human population itself will equal the earth's weight. In other words, another earth will have to be accommodated within this earth, which is totally absurd. This implies that man will run out of space and resources. Thus the present population growth will get more and more unfavourable for human survival as it will destroy the carrying capacity in the long run.

3.2. POPULATION STABILISATION/SUSTAINABLE DEVELOPMENT (INDIAN CONTEXT)

India accounts for 16 per cent of the world population but only 2.4 per cent of the land area. Hence per capita availability of land in the country is 0.48 hectare as against 4.14 hectares in USA, 8.43 hectares in CIS and 0.98 hectares in China. Our population is concentrated in the river valleys (40 per cent in Ganga basin). Man:land ratio in relation to arable land is only 0.27 hectare in India which is quite low for a predominantly agricultural country.

The concept of carrying capacity, as stated above, implies that only a limited number of people can be supported by the resources of a country. Due to demographic (population) pressure, economic pressures build up while cropland, grassland, woodland and fisheries that support the human population by supplying goods and services, are adversely affected. As people multiply, resources obviously shrink. The future population growth has to be related to the resource base in order to have sustainable development. Population stabilisation (i.e. keeping the population level low, preferably below 0.5 per cent) as in the developed countries) is the key to economic development and resource management. This is the success story of all developed countries.

Population explosion is a time-bomb which must be diffused sooner the better. It is one of the major crisis facing the nation. We have to break the vicious P-triangle—Population—Poverty—Pollution on an emergency basis for our own survival.

Kerala Model

Kerala has earned the distinction of lowest birth rate among Indian states since 1960—25 per 1000 as against 33 per 1000, Indian average. It is comparable, as a state, to West Bengal—both are densely populated, both have emphasis on education and both are under similar political system of long periods of rule by left-oriented parties. But the comparison ends here. Kerala is poorer than West Bengal (per capita income, both urban and rural, less than West Bengal), less industrialised than West Bengal and has greater inequalities in income. Yet Kerala's birth rate is lower than that of West Bengal since 1960. This belies the general feeling that greater industrialisation and incomes lead to lower birth rates. The success story of Kerala is explained below.

The reasons behind the low birth in Kerala are—higher age of marriage for women at 21 years (Indian average 17 years); higher female literacy rate 53 per cent (Indian average 13 per cent); greater emphasis on primary education (education budget 60 per cent spent on primary education/West Bengal 38 per cent); better public distribution system of food among 97 per cent of population; better health facilities in rural areas and greater success of family planning programmes. Thus Kerala shows the way to population stabilisation in India.

3.3. WOMEN AND ENVIRONMENT

Women, particularly poor village women, are the worst victims of environmental degradation. Their day starts with a long march in search for fuel, fodder and water. Many such women spend up to 10 hours a day fetching fuel, fodder and water. An average family in Karnataka village walks about 5 km daily on average for collecting fuelwood only, equivalent to walking from Delhi to Bombay every year. In deforested hill and forests the treks are longer and more difficult. By lighting firewood in chullahs, they turn their kitchens into gas chambers where they spend 2-3 hours daily for cooking food. They are exposed to heavy fumes and smoke and suffer as much as a chain smoker smoking 20 packets of cigarettes a day.

Biomass plays a crucial role in meeting daily needs of the vast majority of rural households. The different forms of biomass are food, fish, fuel (firewood, crop wastes and cowdung); fodder, fertiliser for agricultural fields (cowdung, organic manure, forest wastes); building materials (timber and thatch) and medicines (herbs). The country's biomass base is affected by deforestation and loss of vegetation cover. It is the village women who are worst hit by deforestation and devegetation (loss of vegetation cover). Similarly, the forest dwellers, viz. tribals and nomads those who wander from one place to another and have no fixed location, face total social destruction—they are converted into the uprooted lot of "*environmental refugees.*

Woman Power

In the 18th century some 350 Bishnoi men and women from the Bishnoi community in Rajasthan were determined to protect their forest. They clasped the trees when the Maharaja of Marwar sent some wood-cutters along with soldiers to cut down those trees which were required for a brick kiln (furnace) to make slabs for his palace. But the group of men and women resisted by continuing to clasp the trees and finally they were hacked to death. This was the precursor (forerunner) of the Chipko Movement two centuries later (1983).

The women of the Garhwal hill region of UP were at the vanguard of the Chipko Movement (*see* Chapter 10). They organised and executed the movement with utmost sincerity and spread their base all over the Garhwal region. Thus organised woman power prevented further ecological crisis in the Himalayan region.

Another shining example of woman power was the case of Khirakot women in Almora district of UP Himalayas. A Kanpur contractor was setting up soapstone mines in the hills and thereby destroying the ecology of the region. The Khirakot women sprang into action—they got organised, stopped the work of the mines and also fought legal case against the contractor. Finally they won the battle, both legal and environmental and got the contractor's licence cancelled.

3.4. HUMAN DEVELOPMENT

The economic status of a country is normally judged by its Gross Domestic Product—GDP (consumerism, i.e. commodities purchased per year, consumer durables, financial status). But since the nineties the United Nations introduced the concept of Human Development Index—HDI replacing GDP. This is an estimate of human resources development as measured by three parameters—life span, literacy and standard of living. It determines the quality of life in a country. All the countries in the world have been listed in terms of HDI.

The latest United Nations Report (2000) on Human Development and HDI list (Human Development Index) reveals some interesting facts and figures regarding the status of India in the world perspective.

India has moved up by four places from 132 to 128 in the HDI list while Pakistan ranks 135, Bangladesh 146, and Sri Lanka has moved up to position 84. Canada retains the top position (number one) for the seventh consecutive year while Japan has moved down from position number one to eighth during a span of 10 years. The HDI rankings of other countries in the high category are: Norway (2), USA (number 3), Australia (number 4), Sweden (number 5), Belgium (number 6), the Netherlands (number 7), Japan (number 8), Britain (number 9) and Finland (number 10).

According to the Report, even in the richest nations relative prosperity in some countries like USA has failed to improve lives. Although USA has the second highest per capita income among 18 richest nations (OECD, i.e. Overseas Economic Cooperation and Development countries), it has a fairly high poverty rate. The main reason is the prevalence of functional illiteracy (20 per cent).

In general, however, the deprivations cast their dark shadow over most of the world. Some 1.2 billion people live below poverty line, with income less than one dollar a day; more than 1 billion people in the developing countries lack access to safe water and more than 2.4 billion people lack adequate sanitation.

In the HDI list, India occupies 128th position. Per capita GDP (Gross Domestic Product) values for India ($ 300) are less than 1.5 per cent of those of Japan ($ 24,000) and USA ($ 20,000). Inspite of remarkable progress in science and technology, India remains one of the poorest countries in the world, with 50 per cent population living below poverty line, 44 per cent below literacy line and 25 per cent without access to proper health care.

QUESTIONS

1. Trace the world population growth up to 2000 A.D.
2. What is population explosion? How does it affect the economy of a country like India?
3. What is doubling time of population? Explain with reference to world population and Indian population.
4. How is infant mortality related to life expectancy?
5. Write notes on:
   (a) Carrying capacity of the Earth.
   (b) Population stabilisation and sustainable development.
   (c) Human Development Index.

Unit 4

Pressure of Population on India's Environment

4.1. TRENDS IN POPULATION GROWTH

During the British rule in India the population growth rate was insignificant. Agriculture was neglected and peasants subjected to oppression as a matter of the colonial policy which led to poverty, malnutrition, famine, epidemics and death. The birth rate and death rate were almost balanced and this caused population drop by 1.1 per cent during 1911-21. Slow increase in population was noticed from 1931 onwards which got a boost after independence in 1947.

The population of 340 million in 1947 more than doubled to 683 million in 1981 within a period of 34 years and to 1 billion in 2000. In other words, India added a second India within 34 years and a third India within 53 years since Independence. The population has increased by 160.6 million during 1981-91 and after touching the 1-billion mark by May, 2000. India became the second most populous country in the world only next to China. At present India is adding a population of 18 million every year roughly equal to the population of Australia. At this rate it is expected to overtake China by 2025 A.D. and earn the distinction of becoming the most populous country in the world.

The population growth rate in India is shown in the Table 4.1 below.

Table 4.1: Population Growth in India (1901-2000)

Year	Population (in million)	Decade growth rate (%)
1901	238.1	–
1911	252.1	+ 5.75
1921	251.3	– 0.31
1931	279.0	+ 11.00
1941	318.9	+ 14.22
1951	361.1	+ 13.31
1961	439.2	+ 21.51
1971	548.2	+ 24.80
1981	683.8	+ 24.75
1991	846.3	+ 23.51
2000 (May 11)	1000 (1 billion)	

Such fast growth of population (+24%) since 1980 is called Population explosion. This is responsible for rapid environmental degradation.

Urbanisation

Industrial progress along with job opportunities led to growth and mushrooming of cities all over the world as centres of commerce and production and of art and culture. These also encouraged migration of rural population to cities. At the time of independence the share of urban population did not exceed 8 per cent but by 1991 it rose to 36.2 per cent (217 million). The number of million plus cities increased from 3 of pre-independent period to 23 by 1991. The top ten such cities are listed in Table 4.2.

Unplanned growth of urbanisation has led to collapse of infrastructure particularly in metropolitan cities and also apalling living conditions in slum colonies which have become a part of city life with poor quality of life.

Table 4.2: Top Ten Million-plus Cities of India (1991)

	City	Population (in million)
1.	Greater Bombay	12.57
2.	Kolkata	10.92
3.	Delhi	8.38
4.	Chennai	5.36
5.	Hyderabad	4.28
6.	Bangalore	4.09
7.	Ahmedabad	3.30
8.	Pune	2.49
9.	Kanpur	2.11
10.	Nagpur	1.66

4.2. IDENTIFICATION OF HIGH, MEDIUM AND LOW DENSITY AREA IN INDIA

The number of persons per square kilometre is called the *Density of Population*. This gives a measure of the growth of population. The country-wise distribution of population shows uneven trend. Thus according to 1991 Census, W. Bengal shows the highest population density, 767 persons per sq.km, closely followed by Kerala 747 persons per sq. km. while Arunachal has the lowest population density 10 persons per sq.km.

On the basis of the density of population it is possible to identify three major areas—high density area, medium density area and low density area. The vast Gangetic Plain (U.P., Bihar, W.B.), Brahmaputra valley followed by the coastal plains along the Bay of Bengal and the Arabian Sea represent the most densely populated areas in India. In these areas there are navigable rivers, adequate monsoon rainfall, flat low land and fertile soil. These areas include W. Bengal, Bihar, Orissa, U.P., Kerala, Tamil Nadu, Andhra Pradesh, Karnataka, Assam and Tripura. Here the density of population varies from 200 to above 300 persons per sq.km.

The plateau region, Madhya Pradesh, parts of Punjab, Haryana, Rajasthan and Gujarat are areas of medium density of population with 100-200 people per sq.km. This region is not fertile and does not receive adequate rainfall. But there are sufficient irrigation facilities for agricultural purposes.

The low density area has population density less than 100 people per sq. km. The high altitude regions of Kashmir, Ladakh, Himachal Pradesh, the Terai region at the foot hills, Sikkim, hilly states of N.E. India—Manipur, Meghalaya, Mizoram, Nagaland and Arunachal Pradesh (lowest density 10 people per sq. km.) and also the Union Territory of Andaman and Nicobar Islands. The reason for the low population density are hilly terrain, rocky mountain, dense forest, very cold climate and heavy rainfall.

4.3. IMPACT OF POPULATION PRESSURE ON INDIA'S FOOD AND NATURAL RESOURCES

While India occupies 2.4% of world's land area, it accounts for 16% of the world population. Also the hard fact remains that among 1 billion population about 500 million are illiterate and 400 million live below the poverty line. Such huge population has enormous demands for food, shelter, water, fuel etc. which have impact on the natural resources.

Food: Rice is the principal cereal crop which mainly grows in the monsoon-dominated regions of W. Bengal, Assam, Bihar, Orissa, eastern U.P. and southern states of Andhra Pradesh, Tamil Nadu, Kerala and Karnataka. The second and most important crop of India is wheat which grows in Punjab, Haryana, U.P., Madhya Pradesh and in the plains of Gujarat and Rajasthan. The Green Revolution in the sixties/seventies boosted wheat production in Punjab and Haryana. Use of high-yielding variety of wheat (Mexican), improved irrigation facilities, chemical fertilisers and pesticides showed a record production of wheat in Punjab and Haryana. This was also reflected in increased overall food production. From 50 million tons in 1950 the production jumped to 100 million tons in 1970, 180 million tons in 1993 and 200 million tons in 1998. But since the nineties they reached the plateau. The food grain production cannot keep pace with our population explosion. It is well-known that while agricultural production increases arithmetically, population increases geometrically. The cultivated land (about 143 million hectares) has already reached the saturation point of carrying capacity since it has been subjected to land degradation, loss of soil fertility due to intense cultivation, excessive use of fertilisers and pesticides.

Water: The water front presents another gloomy scenario. Heavy withdrawal of ground water for intensive cultivation, for more than two decades caused water tables to fall threatening food production and food scarcity of the country. According to the estimate of the International Water Management Institute (IWMI) withdrawals of ground water are double the rate of aquifier recharge from rainfall. As a result, water tables are falling all over the country. This is likely to reduce India's grain harvest by one fourth.

On the other hand, population explosion has, in general, exerted tremendous pressure on water resources. Our available water resource is 690 cubic kilometre (km^3) surface water and 452 km^3 ground water. This is not adequate to meet the growing demand for water and is likely to result in water scarcity in near future all over.

Soil: Fast growth of human and livestock population has led to continuous degradation of soil due to human activities such as intensive cultivation, overutilisation of marginal grazing land, random deforestation, fuel and fodder collection, road construction, surface mining etc. It is estimated that every year we are losing 6000 tons of top soil resulting in soil erosion and loss in soil fertility.

Forest: Population explosion exerts tremendous pressure on forest resources to meet the ever-increasing demands for land, furniture, fuel etc. This has resulted in reckless deforestation with annual loss of 1.3 million hectares of forests. Deforestation is associated with extinction of valuable species of medicinal plants and poses hazards for man.

Fuel: Population pressure has imposed severe strain on non-renewable and conventional energy resources like fossil fuels (coal, petroleum and natural gas) and also mineral resources such as iron, copper, lead, silver and gold. The increasing consumption of fossil fuel for domestic and industrial needs and transport has generated air pollution and health hazards particularly in cities and industrial areas.

It may be concluded that population, food, natural resources and environment are closely inter-related. As a matter of policy, the population growth has to be reduced from 2.4 per cent to 1 per cent (China 1%, western countries 0.5%) in order to protect the environment and save the next generation.

All our problems of environment are linked to population and poverty. We are trapped inside the vicious *P-triangle*—Population, Poverty and Pollution. Among one billion population, about 50 per cent or 500 million people are illiterate and 400 million people live below the poverty line. According to the Nobel Laureate Professor Amartya Sen, eradication of illiteracy and poverty should be given top priority for pulling up India from the present mess. This will be effective after strong family planning and then only environmental pollution can be arrested.

QUESTIONS

1. What is Population Explosion? Explain with reference to population growth between 1981-1990 and 1991-2000.
2. Name the top ten million plus cities of India and state their population.
3. What are the (a) high population density areas and (b) medium population density areas of India?
4. Discuss how population pressure affects
   (a) Food
   (b) Water
   (c) Forest Resources
5. What is P-Triangle? Suggest how we can solve the P-Triangle.

N.B. For Objective Tests, see Unit-11.

Unit 5

Environmental Consequences of Human Intervention-I

Man occupies the centre of the Biosphere. The primitive man (about 25 million years ago) lived in caves and was panicky about the forces of Nature—rain, thunder, lightning, cyclones and wild animals. In his efforts for self-defence, he discovered fire, small arms made of stones and marched along the path of civilisation. He learnt the art of agriculture 10,000 years ago and the techniques for protection from wild animals and natural hazards. In course of time he acquired the science and technology of controlling the natural barriers but at the same time he lost his eco-friendly relation with his environment leading to its degradation. This happened after the Industrial Revolution (1780). Two centuries of progress in Science and Technology, accompanied by rapid industrial development, caused environmental disaster, threatening survival of man himself on earth.

Some of the adverse effects on environment as a result of human intervention will be discussed in this chapter.

5.1. GREENHOUSE EFFECT (GLOBAL WARMING)

Carbon dioxide is a non-pollutant gas in the atmosphere and a minor constituent (356 parts per million) but it is of serious concern for the environment for its ability to change the global climate.

The earth's surface partly absorbs sun's rays while emits long-wave infra-red radiation (8000-25000 nanometres; 1 nanometre = 10^{-9} metre = 1 nm). Carbon dioxide and water vapour in the atmosphere strongly absorb infra-red radiation (14,000-25,000 nm) and effectively block a large fraction of the earth's emitted radiation. The radiation thus absorbed by carbon dioxide and water vapour is partly returned to the earth's surface. The net result is that the earth's surface gets heated and the phenomenon is known as the *Greenhouse Effect* (Fig. 5.1).

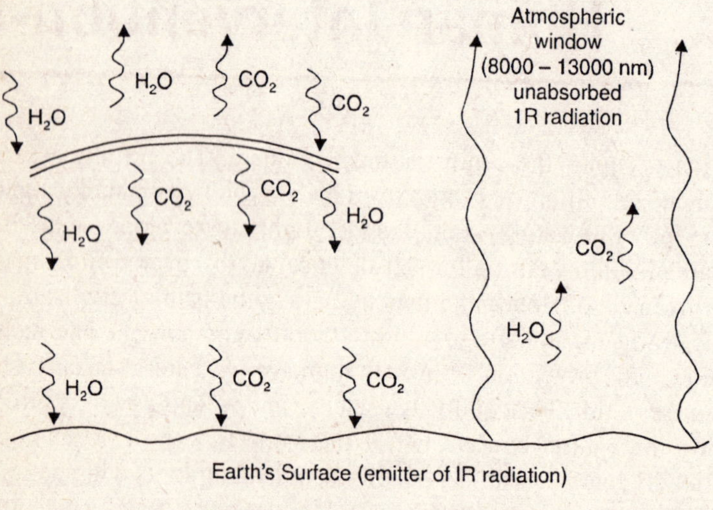

Fig. 5.1. The Greenhouse Effect

The carbon dioxide level in air has increased from 280 ppm (pre-industrial revolution era 1780) to 350 ppm at present in two centuries. Fossil fuel (petrol, diesel, coal) combustion is the major source of increase of carbon dioxide level increasing at the rate of 1-2 per cent per year. At this rate of increase, the earth's surface temperature may rise as much as 2°C in the next 100 years. However, nature has its check and balance system. The rate of increase of carbon dioxide is only 50 per cent of its expected magnitude due to the sinks, viz. oceans and photosynthesis by green plants (Fig. 5.2).

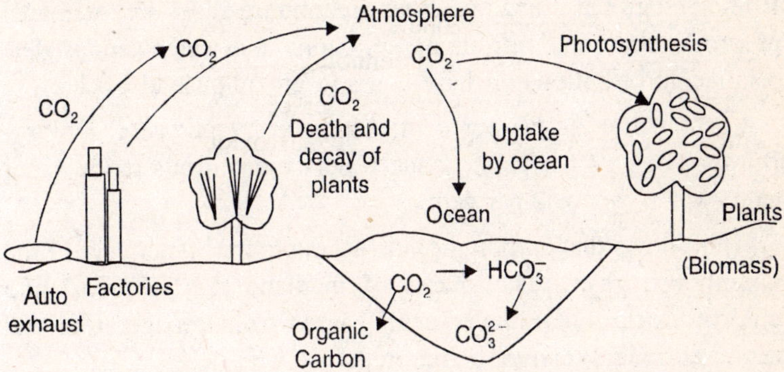

Fig. 5.2. Sources and Sinks of Carbon dioxide

It may be noted that a slight rise in temperature even by 1°C, can have adverse effect on the world food production. (1) Thus wheat producing zones in the northern latitude will be shifted from CIS (former USSR) and Canada to the north pole and in India from UP, Punjab, Haryana to the Himalayas, i.e. from fertile to non-productive soil. In other words, wheat production will badly suffer. (2) The biological productivity of the ocean will fall due to warming of the surface layer. This reduces transport of nutrient from deeper layers of the ocean to the surface by vertical circulation. Moreover, there will be less photosynthesis by marine plants. In other words, the production of sea food (marine plants and fish) will decline. Sea food constitutes more than 30 per cent of our food supply. (3) Another disastrous effect is the rise in sea levels by as much as 15 cm in the next 100 years due to partial melting of polar ice caps. This sea level rise would threaten coastal cities (Kolkata, Mumbai, Chennai, etc.) and some 60-odd island nations such as Maldives, Bangladesh, etc. which will be drowned under the sea. When Himalayan snow melts and gets exhausted, the Himalayan rivers including Ganga may dry up—the Ganga valley will be hot and north India will lose its population base. At the same time India will lose its major life line, Ganga.

Other Greenhouse Gases

Carbon dioxide is not the only culprit responsible for Greenhouse Effect and global warming. Other greenhouse gases are: methane, chloroflurocarbons, nitrous oxide, ozone and water vapour. The relative contributions of these gases to greenhouse effect are:

Carbon dioxide 50 per cent; Methane 19 per cent; Chloroflurocarbons 17 per cent; ozone 8 per cent; nitrous oxide 4 per cent; water vapour 2 per cent.

This shows that carbon dioxide accounts for 50 per cent of the greenhouse gases. The shares of methane (19 per cent) and chloroflurocarbons (17 per cent) (gases from refrigerators and air-conditioners) cannot be ignored.

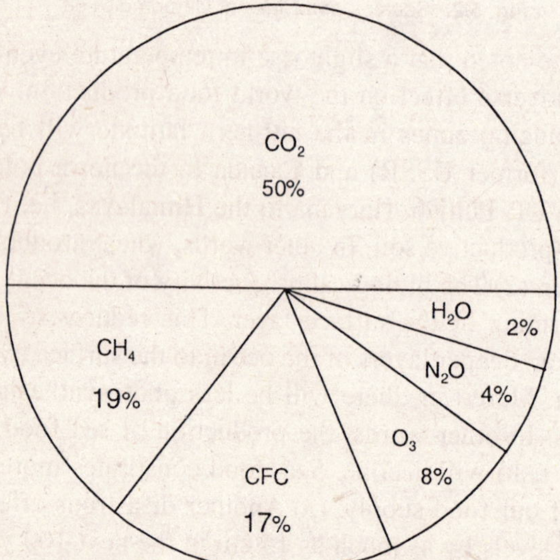

Fig. 5.3. Greenhouse Gases (relative shares)

5.2. OZONE HOLE

In the stratosphere, the second region of the atmosphere, ozone is present in small quantities but it is protective shield for the earth. Ozone strongly absorbs ultraviolet radiation from the sun

(295-320 nm) which is injurious for life on earth. Thus it protects living species on earth. But recent human activities have injected some dangerous chemicals in the stratosphere which consume ozone and reduce its concentration. This is the phenomenon of *ozone hole* in the stratosphere.

Exhaust gases from jet aircrafts and artificial satellites discharge nitric oxide (NO), nitrogen dioxide (NO_2), etc. which immediately react with ozone.

$$\underset{\text{(ozone)}}{O_3} + \underset{\text{(nitric oxide)}}{NO} \longrightarrow \underset{\text{(oxygen)}}{O_2} + \underset{\text{(nitrogen dioxide)}}{NO_2}$$

$$O_3 + NO_2 \longrightarrow O_2 + \underset{\text{(nitrogen trioxide)}}{NO_3}$$

Chloroflurocarbons (CFC) are used as coolants in refrigerators and air-conditioners. These slowly pass from troposphere and stratosphere and once there, they stay for 100 years.

In presence of ultraviolet radiation (200 nm) from the sun, CFC breaks up into chlorine-free radical (Cl) which readily consumes ozone.

$$CFCl_3 \xrightarrow[\substack{\text{ultra-violet} \\ \text{radiation}}]{h\nu} CFCl_2 + Cl^-$$

$$Cl^{\bullet} + O_3 \longrightarrow ClO^{\bullet} + O_2$$

$$ClO^{\bullet} + O \longrightarrow Cl^{\bullet} + O_2$$

The free radical ($Cl^{\bullet}$) is regenerated and continues the chain reaction. It is estimated that one molecule of CFC consumes one lakh molecules of ozone. The damage by CFC continues for 100 years. Even if CFC production is stopped now all over the world, the CFC that is already there in the stratosphere will continue to damage the ozone layer for the next 100 years.

In 1979 ozone hole was observed in the sky over Antarctica where ozone layer thickness was reduced by 30 per cent. Later on ozone hole was discovered in the sky over the thickly populated northern hemisphere. Here in winter ozone thickness was reduced by 4 per cent and in summer by 1 per cent.

Ozone hole allows passage of ultraviolet radiation to the earth where it causes skin cancer, eyesight defect, genetic disorder, etc. in the biosphere (man, animal and plant). In Europe and USA there is an increase in the cases of skin cancer among people while some million people are suffering from eye cataract.

In Montreal Conference (1987) and London Conference (1992) it was decided that the developed countries would totally ban CFC production by 2000 and the developing countries by 2010 AD. But as stated above, even after the ban is enforced, the CFC and Cl• shall continue their havoc for another 100 years. Researches are on for development of CFC-substitutes as coolants for refrigerators and air-conditioners.

El Nino

Greenhouse effect, as discussed above, is an atmospheric warming phenomenon while El Nino is an ocean warming phenomenon. El Nino is a Spanish term for the Christ child as it starts in December in South American coast but it affects climate over half the globe.

Under normal conditions the water of the Eastern Pacific off Ecuador, Peru and Chile is cooler by 10°C than the waters of the Western Pacific. This promotes fisheries along the Eastern Pacific coast since cold waters, rich in nutrients, spring from the deep layers of the ocean. The trade winds blow along the equator from Southeast Pacific towards West pushing warm waters out to the ocean. But once every three to five years the trade winds reverse their direction, i.e. from west to east thereby pushing warm waters to the east. As a result, the Eastern Pacific water warms up to about 4°C (24° to 28°C) which disrupts the anchovy fishery, key to the Peruvian economy. This happens when El Nino appears. It not only kills fish but warms the air which lowers the atmospheric pressure and sparks storms and heavy rainfall along Chile coast and spreading up to Southern California (USA).

In 1982, El Nino raised sandstorms in Australia, cyclones in Tahiti, caused droughts in Africa and floods along the Californian coast in USA. Recently in 1998, El Nino activity has been felt all over the world. In the east coast of USA, winter was missing in January-February and in the west coast there was record rainfall in 100 years of US history. Research is on to understand the science of El Nino so that it can be forecast and advance warning signal sent to the regions where lives and properties can be saved in time.

5.3. DEFORESTATION

(i) **Role of Forests:** Forests are valuable renewable natural resources for any country. They play the most important role in maintaining the ecological balance in the environment.

Forests conduct different functions in different regions. They regulate water supply, feed springs and streams in hilly areas while in the plains they help drainage and reduce waterlogging and swampiness in wet areas. They slow the flow of rainwater and thus limit floods, soil erosion and landslides.

India has a rich tradition of forest culture ("Aranya Sanskriti") dating back to the vedic era (about 3000 B.C.). The basic concepts of environmental ethics were outlined in the Vedas and Upanishadas and taught in Gurukul system of Education. The original forest cover of 80% in the Vedic era has now dwindled to less than 20%. Indian forests are under heavy pressure with less than 2% of global forest area—they cater to 16% of global population and 16% of global cattle population (in India).

(ii) **Material benefits from Forests:** Forests offer multiple benefits to man, animals, birds and the ecosystem in general. According to tribals who live in forests, forests provide them with *three essential 'F's*—Fuel, Food and Fodder. They are the sources of fruits, nuts, vegetables, honey, spices, animal products, fuel, timber, wood, medicines and also raw materials such as oil, resin, gum, rubber, wax, fibre, bamboo etc. It is the tribals who are worst hit with deforestation, as they have to leave their forest homes, lose their identity and are reduced to the status of *Environmental Refugees* in human society.

(iii) **Causes of Deforestation:** By deforestation is meant permanent removal of forest and its under growth. Population explosion is responsible for enormous pressure on forest resources leading to over utilisation of forest land, wood and non-wood forest products etc. for various purposes.

Forests are cleared for agriculture, farming, plantation of cash crops, housing, road construction etc. Rapid industrialization, urbanization, construction of railways for communication have brought about large scale deforestation. During the period of Roman empire Europe had 90% forest cover while at present it is hardly 30%. The same scenario is observed in the Far East and Indian sub-continent. The Thar desert of Rajasthan in India as we see today was the home of dense jungle 2000 years ago. Deforestation for agricultural land and housing has transformed it into a desert.

The forests in the Mediterranean region and China have also been devastated by human activities. In the Mediterranean Europe at present forests occupy only 5% of the original area. The Chinese have been known to be the worst destroyer of forests for thousands of years. The world's last great virgin forest in the Amazon basin (S. America) is today facing extinction. Much deforestation has also occurred in Brazil due to coffee plantation.

The tea plantation in Upper Assam and coffee plantation in the Nilgiri hills has wiped out the forests in these areas. The Shivalik range of forests below the Himalayas in North India had dense forest cover which is now in semi-desert conditions due to over-exploitation of sal trees for commercial and industrial purposes.

Among other major causes for deforestation in tropical countries may be mentioned conversion of jungles into pasture lands for overgrazing by sheep and cattle (Latin America), reckless felling of trees for timber and fuel, practice of shifting cultivation (Jhum cultivation) and forest fire. Overgrazing by sheep and cattle turn pasture lands into deserts. Timbers are used for making furniture and housing and also for industrial purposes. Demands for fuel among rural people and tribals have been met by deforestation.

In the hilly regions of North Eastern India the old practice of "*slash and burn*" i.e. *Jhum Cultivation* deserves special mention. This is a traditional process of deforestation. Trees and bush in a forest area are first cut and burnt for cultivation and after two or three years when the crop yield decreases, the farmers change their old sites and shift to new areas. The practice is also in vogue among the tribals in Orissa, Madhya Pradesh and Andhra Pradesh.

Loss of forests by fire incidences from spark by man-made activities are of common occurrence in Garo hills, Khasi hills, Arunachal Pradesh, Nagaland and Assam in N.E. India and also in the Himalayas in U.P., Himachal Pradesh. Prolonged dry spell in tropical regions accelerates consumption of forests by fires. Corniferous forests in hill areas are more inflammable than others and easily become victims of devastation.

Effects of Deforestation: The environmental damages due to deforestation are enormous.

(i) soil erosion causing drought condition and favouring natural calamities like flood and cyclone as annual events;

(ii) extinction of important species including medicinal plants and fauna, affecting biodiversity;

(iii) increase in carbon dioxide content due to forest scarcity ensuring global warming;

(iv) lower stock of aquifier (underground water) due to vanishing forests.

In short, the quality of life on earth badly suffers due to environmental degradation resulting from deforestation.

5.4. DESERTIFICATION

The twin evils of deforestation and land degradation bring in their trail desert-like conditions leading to desertification. Natural and man-made factors contribute to desertification. Climatic changes leading to temperature rise, less rainfall, wind velocity and soil erosion cause drought in dry, sub-humid areas. Prolonged droughts over several years favour desert-like conditions. Increased dust particles in air and loss of vegetation cover over degraded land expedite the process of desertification.

Man-made factors are, of course, desertification due to human and livestock population explosion. Indiscriminate felling of trees, overgrazing by livestock, shifting cultivation, extensive mining, wind, soil erosion etc. badly degrade the ecosystem. Field rodents also have substantial contribution to desertification.

Desertification strikes at the root of the resource base so that the renewable resources get lost. The worst victims are Africans living in Sahel region and Somalia. They fall prey to famine and starvation deaths. Desertification can be checked and prevented by taking adequate safeguards against desertification, prevention of overgrazing, control of soil erosion and proper water and land management.

For *Questions see* at the end of Unit 6.

Unit 6

Environmental Consequences of Human Intervention-II

The various segments of the environment—air, water, soil and biosphere—are in dynamic equilibrium. Any addition of extraneous material (solid, liquid, gas) will disturb this equilibrium and cause change in physico-chemical or biological characteristics of air, water and soil which is harmful to the living organisms including man. This is the phenomenon of *pollution*. It has two sources—natural and man-made (anthropogenic). Thus emission of gases and particles from volcanoes is an example of natural pollution while automobile exhausts are sources of man-made pollution. Pollutants are substances, organic or inorganic, which cause pollution. Pure or unpolluted water is colourless, odourless and tasteless but polluted water has colour, odour and taste. Similarly, polluted air has colour and foul odour.

According to the Environment Protection Act 1986, "*Environmental pollution* can be defined as any adverse change in the environment caused directly or indirectly by man in the process of his living." However, this has omitted natural pollution probably out of its urge to emphasize man-made pollution only.

This chapter deals with different types of pollution: Air pollution, Water pollution, Soil pollution, Radiation pollution and Noise pollution and their control methods.

6.1. WATER POLLUTION

The normal uses of water for public supply are—recreation (swimming, boating, etc.), fish, other aquatic life and wild life, agriculture (irrigation), industry, navigation, etc. Any change in the dynamic equilibrium in aquatic ecosystem (water body/biosphere/atmosphere) disturbs the normal function and properties of pure water and gives rise to the phenomenon of water pollution. The symptoms of water pollution of any waterbody/ground water are:

- Bad taste of drinking water,
- Offensive smells from lakes, rivers and ocean beaches,
- Unchecked growth of aquatic weeds in water bodies (eutrophication),
- Dead fish floating on water surface in river, lake, etc.
- Oil and grease floating on water surface.

The quality of water is of vital concern for mankind since it is directly linked with human welfare. It is a historical fact that faecal (human excreta or stool) pollution of drinking water caused water-borne diseases which wiped out entire populations of cities. In the developing countries like India, every day some 25,000 people die of water-borne diseases, e.g. jaundice, hepatitis, cholera, dysentery, etc. In India about 2 lakhs out of 6 lakh villages have no access to safe drinking water—women have to walk 1-14 km daily for collecting water for drinking and cooking. In urban areas 40 per cent people are without access to safe water. The major sources of water pollution are domestic sewage from urban and rural areas, agricultural runoff (wash water) and industrial waste which are directly or indirectly discharged into water bodies.

Water Pollutants

The large number of water pollutants are broadly classified under the categories:

1. Organic pollutants,
2. Inorganic pollutants,
3. Sediments,
4. Radioactive materials, and
5. Thermal pollutants.

Organic Pollutants

These include domestic sewage, pesticides, synthetic organic compounds, plant nutrients (from agricultural run-off), oil, wastes from food processing plants, paper mills and tanneries, etc. These reduce dissolved oxygen (D.O.) in water. Dissolved oxygen (D.O.) is essential for aquatic life, the optimum level being 4-6 ppm (parts per million). Decrease in D.O. value is an indicator of water pollution. The organic pollutants consume D.O. through the action of bacteria present in water.

Sewage and agricultural run-off provide plant nutrients in water giving rise to the biological process know as *eutrophication*. Large input of fertiliser and nutrients from these sources leads to enormous growth of aquatic weeds which gradually cover the entire waterbody (*algal bloom*). This disturbs the normal uses of water as the water body loses its D.O. and ends up in a deep pool of water where fish cannot survive.

The production of synthetic organic chemicals (more than 60 million tons each year since 1980) multiplied more than 10 times since 1950. These include fuels, plastic fibres, solvents, detergents, paints, food additive, pharmaceuticals, etc. Their presence in water gives objectionable and offensive tastes, odour and colours to fish and aquatic plants.

Oil pollution of the seas has increased over the years, due to increased traffic of oil tankers in the seas causing oil spill and also due to oil losses during off-shore drilling. Oil pollution reduces light transmission through surface water and hence reduces photosynthesis by marine plants, decreases D.O. in water causing damage to marine life (plants, fish, etc.) and also contaminates sea food which enters the human food chain.

Pesticides have been largely used for killing pests and insects harmful for crops and thereby boosting the crop production. At present there are more than 10,000 different pesticides in use. They include insecticides (for killing insects), e.g. DDT (dichloro diphenyl trichloromethane), herbicides (for killing weeds and undesirable vegetation) and fungicides (for killing fungi and checking plant disease).

It has been found that pesticide residues contaminate crops and then enter the food chain of birds, mammals and human beings. The persistent pesticide, viz, DDT (which is not degraded in the environment) accumulates in food chain, getting magnified in each step from sea weed to fish and then to man by about ten thousand times (10^4). Thus the average level of DDT in human tissues is found to be 5-10 ppm, maximum being among Indians (25 ppm) compared to Americans (8 ppm).

Inorganic Pollutants

This group consists of inorganic salts, mineral acids, metals, trace elements, detergents, etc.

Acid mine drainage: Coal mines, particularly those which have been abandoned, discharge acid (sulphuric acid) and also ferric hydroxide into local streams through seepage. The acid on entering the waterbody destroys its aquatic life (plants, fish, etc.)

Sediments

Soil erosion, as a matter of natural process, generates sediments in water. Solid loadings in natural water are about 700 times as large as the solid loading from sewage discharge. Soil erosion is enhanced 5-10 times due to agricultural and 100 times due to construction activities. Bottom sediments in aquatic bodies (streams, lakes, estuaries, oceans) are important reservoirs of inorganic and organic matter, particularly trace metals, e.g. chromium, copper, nickel, manganese and molybdenum.

Radioactive Materials

Radioactive pollution is caused by mining and processing of radioactive ores to produce radioactive substances, use of radioactive materials in nuclear power plants, use of radioactive isotopes in medical, industrial and research institutes and nuclear tests. The discharge of radioactive wastes into water and sewer systems is likely to create problems in future.

Thermal Pollutants

Coal-fired or nuclear fuel-fired thermal power plants are sources of *thermal pollution*. The hot water from these plants is dumped as waste into nearby lake or river where its temperature rises by about 10°C. This has harmful effect on aquatic life in the water body whose D.O. is reduced and as a result, fish kill is quite common.

Ground Water Pollution/Arsenic Contamination

Ground water is relatively free from surface contamination as it is located more than about 50 ft. below the land surface and the surface water gets filtered or screened by the underlying layers of soil, sand and stone pieces. But even then it gets contaminated due to leaching of minerals below the earth's surface.

An important case is that of *Arsenic (As)—contamination of ground water*. This arises from excessive pumping of ground water by shallow tube wells for irrigation in some West Bengal districts along the Hooghly river course and also in Bangladesh along the Padma river course. In this process air (oxygen) is injected into ground water bed which leaches the overlying mineral, iron pyrites (iron, arsenic, sulphide) oxidises it and releases arsenic into ground water.

More than one million people in six districts of West Bengal drink arsenic-contaminated ground water from tube wells in the region. Among them 20 lakh people suffer from various diseases related to arsenic poisoning like loss of hair, brittle nails, bronchitis, gangrene, etc. Several hundred deaths have also been reported.

Similar calamity has threatened the lives in Bangladesh in the districts along the Padma river course.

Case Study of Ganga Pollution

The most typical example of river pollution is the *Ganga Pollution.*

The Ganga originates from the Himalayan glacier and flows along a stretch of some 2525 km before joining the Bay of Bengal. The Ganga basin is fertile and home of about 40 per cent of population (400 million people) of the country. The river has been hailed as the "Holy Ganga" and regarded as the lifeline of the country. But in recent years it is ranked as the most polluted river of India and a killer in the highly polluted areas.

The Ganga basin carries wash water from 25 per cent of land. It is responsible for agricultural prosperity of U.P, M.P, Haryana, Rajasthan, Himachal Pradesh, Bihar and West Bengal. Ganga is the source of drinking water in the region and irrigation water for agriculture—she also supplies fish to the local markets and water to industries on both sides of the river. The Ganga basin provides maximum population density—many class I (population 100,000 and above), class II (pop. between 50,000 and less than 100,000) and class III (pop. 20,000 to less than 50,000) cities have grown in this region. Both domestic and industrial sewage join the Ganga river without any treatment and thus cause terrible pollution.

Hooghly river (in West Bengal) near Calcutta presents the worst polluted zone. There are more than 150 industries on both sides of the 125 km—stretch river belt—there are about 270 outlets of untreated sewage to the river Hooghly. The entire 140 sq. km metropolitan area covering both banks of the Hooghly river is exposed to ecological disaster. Besides huge quantities of soil from soil erosion due to extensive deforestation are washed by rain water into the river causing siltation. This reduces the flow of water in the Bhagirathi-Hooghly river with the result that ultimately the river will be choked and dead. In 1919 the flow of water in the Ganga was 1,10,000 cusecs (1 cusec = 1 cubic foot of water flowing per second) whereas in 1971 it was 40,000 cusecs only which during summer drops to 20,000 cusecs. This should be enough to sound the alarm bell to the Government—Calcutta and Haldia ports can survive only on 40,000 cusec of water.

6.2. WASTE WATER TREATMENT

Water pollution is caused by domestic sewage (84 per cent) and industrial sewage (16 per cent). Though the latter has less load on water body, it contains toxic matter (inorganic and organic) which are more hazardous.

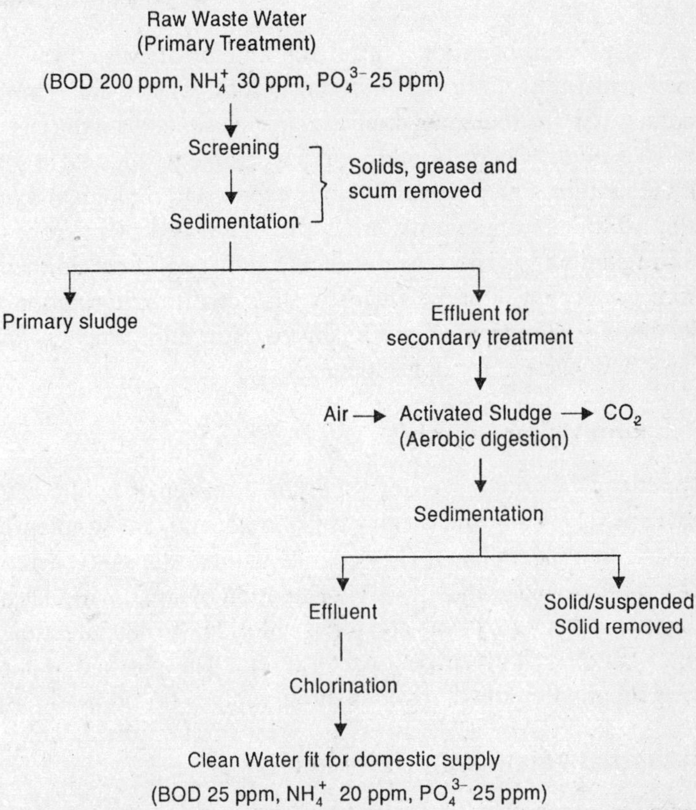

BOD = Biological Oxygen Demand (Index for organic matter content)
1ppm = 1 part per million i.e. 1 in 10^6 parts

Fig. 6.1. Municipal Waste Water Treatment (Primary and Secondary)

Domestic Waste Water Treatment

Sewage treatment plants, in general, depend on biological decomposition of non-toxic organic wastes using bacteria. Such biological decomposition is carried out under aerobic conditions, i.e. in presence of plenty of oxygen.

Organic wastes (carbon, oxygen, hydrogen, sulphur fluorine)	+	O_2 (oxygen)	$\xrightarrow{\text{bacteria}}$	CO_2 (carbon dioxide) H_2O (water) + NO_3^- (nitrate) + NH_4^+ (ammonium ion)

The process commonly used for municipal waste water, in shown in Fig. 6.1. In the first stage, solid wastes are removed from water by screening—any scum (suspended matter) is removed and the sludge (muddy solid or sediment) allowed to settle at the bottom. The residual liquid is exposed to biological oxidation of soluble organic materials through a bed of microbes in activated sludge. Then the solids are removed after sedimentation. Finally the liquid effluent is subjected to chlorination for destroying pathogenic micro-organisms. Now this effluent is fairly clean and suitable for domestic use.

Drinking Water Supply

Treatment of drinking water supply is a matter of public health concern. The water treatment plants, in general, are simpler than sewage treatment plants. They operate in three steps—(i) aeration to settle suspended matter; (ii) coagulation of small particles and suspended matter by lime and ferric chloride; (iii) disinfection by chlorination to kill viruses, bacteria, etc. The purified water is then supplied by municipalities through pipes for domestic uses.

Industrial Waste Water Treatment

Industrial wastes contain toxic chemicals which can damage environment (water, soil, air) much more than domestic sewage. These waste liquids (effluents) can be purified by filtration using activated charcoal or ion exchange resins. Activated charcoal has large surface area and is an effective filter medium for adsorption of organic molecules. Synthetic organic ion exchange resins are very useful for removal of industrial waste metals (cations) and non-metals (anions).

6.3. WATER QUALITY

It is essential to enforce water quality standards in the interest of public health. All developed countries strictly conform to water quality standards. Polluted water generates water-borne diseases which kill millions of people every year all over the world,

Table 6.1: Water Quality Parameters (Domestic Water Supplies) USPH Standards (Upper limits)

Parameters	*(in ppm, except for pH)*
pH	6.0-8.5
Dissolved oxygen (D.O.)	4.0-6.0 ppm
Total dissolved solid	500.0
Suspended solid	5.0
Chloride	250.0
Sulphate	250.0
Cyanide	0.05
Nitrate + Nitrite	10.0
Ammonia	0.5
Calcium	100.0
Magnesium	30.0
Iron	0.3
Lead	0.05
Mercury	0.002
Arsenic	0.05
Chromium (VI)	0.05
Zinc	5.5
Phenol	0.001
Chemical Oxygen Demand (COD)	4.0

All parameters except pH are in ppm. (parts per million)
1 ppm = 1 in 10^6 parts

particularly in developing countries. The United States Public Health (USPH) has laid down standards for water quality parameters (indicators) for drinking water. These are the upper limits in parts per million (ppm) (1 ppm = 1 in 10^6 parts).

The parameters for surface water (rivers, lakes, etc.) are 4-5 times higher than the above values for drinking water.

6.4. WATER POLLUTION CONTROL AND WATER MANAGEMENT—WATER RECYCLING

Clean water is essential for healthy environment to support life systems on this planet. The task of delicately balancing the ratio of available and exploitable water resources and sustaining their quality is most important for India as rainfall distribution is confined to 3-4 months in a year. Moreover, man-made global and local climatic distortions due to global warming (*see* Chapter 6) deforestation, loss of top soil, etc. have adverse effect on the monsoon pattern in India.

India is blessed with good rainfall (average 200 cm in a year) but 70 per cent of it is wasted. The country faces recurring problems of floods, and droughts and highly polluted water resources. It is necessary to do rain harvesting, i.e. build large tanks and reservoirs all over the country to store rain water, flood water and excess water from the Ganga, Brahmaputra and other rivers. The rivers, the lifelines of our culture and economy, are dying because of severe pollution. This water pollution abatement and resource management should be at the top of our national agenda.

The river water pollution has three dimensions—agricultural run-off, industrial effluents and domestic sewage. A typical example is that of the Ganga pollution. The total quantity of fresh water used in the Ganga basin is 150 billion cubic meters out of which 26 per cent is discharged as waste water. Agricultural run-off is 27 billion cubic metre compared to 1530 million cubic metre from industrial and domestic sewage.

Ganga Action Plan

This best illustrates the strategy of water pollution control and management. In order to rejuvenate (i.e. bring back to life) Ganga, the first step is to stop dumping of all sorts of garbage to the river and at the same time to promote public awareness on the issue. The sources of such garbage are domestic sewage, industrial sewage and agricultural run-off. *Every day about 1200 million litres of waste water (mainly from West Bengal) enter the Ganga.*

Industrial sewage has to be treated at the plant site through treatment plants and then the treated effluent discharged into the river or recycled for reuse. Domestic sewage in rural and urban areas should be diverted into separate diversion canals and collected in lagoons, subjected to treatment plants and the resulting clean water used for irrigation in agricultural fields. The biogas (methane, CH_4) emitted from the treatment plants can be used for domestic fuel in industrial areas.

With this objective, the Government of India launched Ganga Action Plan in 1986 and entrusted the job to the State Governments concerned (UP, Bihar and West Bengal) for implementation. The Action Plan has the following agenda:

(i) To divert sewage from the small and large towns on both embankments of the Ganga into separate diversion canal, accumulate it in big lagoons, subject it to sewage treatment plant and use the clean sewage water for agriculture. The emitted biogas (methane, CH_4) from the plant is utilised as domestic fuel through piped gas supply in adjoining towns.

(ii) To construct low cost sanitary latrines (about 30,000) so that people don't commit nuisance on the river embankments.

(iii) To construct the Ganga embankment ghats and construct electric furnaces at selected sites for funeral rites so that the Ganga is saved from dead bodies.

(iv) To make it mandatory for all industries to install treatment plants in their premises for treatment of their solid and liquid effluent as well as gaseous emission.

For the success of the Ganga Actions Plan, it is essential that the public must be involved—at least they must be made aware of the critical condition of Ganga and need for recovery. For this purpose public awareness campaign deserves top priority. It must be noted that Ganga has three major problems—pollution, erosion and siltation. 70 per cent of pollution is generated from towns (class I 27, class II 23 and class II 48) on both sides of the river, 20 per cent from factories and 10 per cent from agricultural lands. The Ganga Action Plan, however, could not be properly implemented. It required active government—people interaction and participation and furthermore, continuous monitoring of the Projects by experts on a long-term basis. The Ganga Action Plan should be executed as a matter of 20 year plan instead of one or two 5 year plans only.

Desalination

Only 0.69 per cent of total water resources is available as fresh water for our consumption. But this resource is shrinking due to reckless water pollution so that in future we may lose it. In that case we have to turn to the vast resource, i.e. sea water (97 per cent), which, at present, is unfit for our consumption due to high salt content. It is possible to utilise sea water with appropriate technology after removal of salts and thus solve water crisis. The technology of removing salts from sea water is known as *desalination*. Israel and Arab countries have started applying this technology for solving their chronic problem of water scarcity.

There are several methods for desalinisation of sea water. The most economical method is to use solar energy for evaporation of sea water and separation of salts. For this purpose metal tanks with low height are used for collecting water. The tanks are painted black on the outside walls so that they become hot in the sun while the tanks are covered on the top with transparent plastic sheets. The water vapour from evaporation of sea water by the sun collects on the inside wall of the plastic sheet—on condensation, this water vapour gives pure water. At the end of the operation, salts are recovered from the tanks. Recently, Israel is getting each year 30 million cubic metres of salt-free sea water.

Electrodyalysis and Membrane Technology are being used in

Japan to obtain salt-free sea water and salts separately. Though these techniques are costly, at present, but with enhanced technology, it may be possible in future to run these on commercial basis.

6.5. AIR POLLUTION

Pure air is colourless and odourless. But various pollutants from natural and man-made sources are entering the atmosphere daily and these disturb the dynamic equilibrium in the atmosphere. This leads to air pollution when the normal properties of air are upset and both man and environment suffer.

Natural sources of air pollution are:

- Volcanic activity, vegetation decay, forest fires emitting carbon monoxide, sulphur dioxide and hydrogen sulphide and tiny particles of solids or liquids sprayed from the seas and land by wind.

Man-made sources are:

- Gases, mists, particulates and aerosols emitted by industries and other chemical and biological processes used by man.

Primary Pollutants

There are five primary pollutants which together contribute more than 90 per cent of global air pollution.

Carbon monoxide, CO

Nitrogen oxides, NO_X

Hydrocarbons, HC

Sulphur oxides, SO_X and

Particulates

Transportation accounts for more than 46 per cent of the total pollutants produced per year and hence remains the principal source of air pollution. Carbon monoxide is the major industrial pollutant, with a tonnage matching that of all other pollutants together. However, particulate pollutants, though minor, are the most dangerous among the primary pollutants (100 times more harmful than carbon monoxide).

Table 6.2: Primary air pollutant sources and their quantities (million tons per year)

Sources	Weight of pollutants produced				Particulates		Total weight of pollutant produced
	CO	NO_x	HC	SO_x	<20μ	>30μ	
Transportation	70	10	10.8	0.8	1.2	1.0	94
Fuel combustion (stationary sources)	1.2	11.8	1.4	21.9	4.6	1.3	42.2
Industrial processes	7.8	0.7	9.4	4.1	6.3	2.7	31.0
Solid waste disposal	7.8	0.8	1.6	0.1	1.1	–	11.2
Miscellaneous	8.5	0.4	6.3	0.1	1.3	–	16.6
Total weight of pollutant produced (in million tons)	95.0	23.6	29.5	27.0	–	19.5	194.6

1μ = 10^{-6} metre (1 part in 1 million parts of 1 metre)

The above data are taken from those in USA (1990). As a matter of fact, USA and other developed countries contribute most to air pollution.

Carbon Monoxide, CO

It is a colourless, odourless and tasteless gas which is injurious to our health. Each year 350 million tons of CO (275 million tons from human sources and 75 million tons from natural sources) are emitted all over the world in which USA alone shares 100 million tons. Transportation accounts for 70 per cent of CO emission. That is to say, diesel and petroleum engines in automobiles are primarily responsible for about 70 per cent of CO emissions.

The sources of carbon monoxide. CO are the chemical reactions:

(i) incomplete combustion of fuel or carbon containing compounds:

$$\underset{\text{(carbon)}}{2C} + \underset{\text{(oxygen)}}{O_2} \longrightarrow \underset{\text{(carbon monoxide)}}{2CO}$$

(ii) reaction of carbon dioxide and carbon-containing materials at elevated temperatures in industries, e.g. in blast furnaces:

$$\underset{\text{(carbon dioxide)}}{CO_2} + \underset{\text{(carbon)}}{C} \longrightarrow 2CO$$

(iii) dissociation of carbon dioxide at high temperatures:

$$CO_2 \rightleftharpoons CO + O$$

Sinks

Part of Carbon monoxide is lost in the upper atmosphere. The major sink is soil micro-organisms. A potting soil sample weighing 28 kg can completely remove in 3 hours 120 ppm carbon monoxide from ambient air. The same soil sample on sterilisation failed to remove carbon monoxide from air.

Control of CO Pollution

The petroleum and diesel-fed automobiles account for major share of carbon monoxide emission. Hence efforts for carbon monoxide pollution control are mainly aimed at automobiles. Use of catalytic converters in the internal combustion engines of automobiles helps in cleaning up the exhaust emissions. Such converters built into the automobile engines promote oxidation-reduction cycles and ensure complete combustion of carbon monoxide, nitrogen oxides and hydrocarbons. The following figure and flow-sheet illustrate the action of catalytic converters. Use of catalytic converters in two stages helps in elimination of pollutants from exhaust gases before they are discharged into the atmosphere.

In the first converter nitrogen oxides are reduced to nitrogen (+ ammonia in presence of finely divided catalyst platinum, and the reducing gases, carbon monoxide and hydrocarbons. The production of ammonia is kept at a minimum under carefully controlled conditions. In the second converter, air is introduced to provide an oxidizing atmosphere for complete oxidation of carbon monoxide and hydrocarbon into carbon dioxide and water in presence of finely divided platinum catalyst.

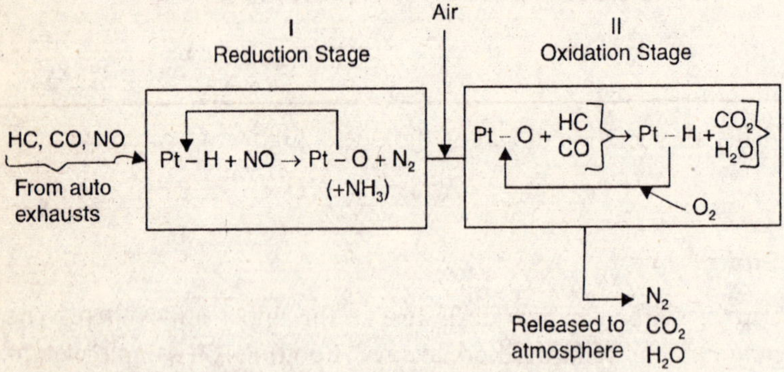

Fig. 6.2. Catalytic converters for treating auto emissions

Thus by means of platinum catalytic converters, auto exhaust emissions are cleaned up through reduction-oxidation reactions. In all developed countries it is mandatory by law for all automobiles to fit their engines with catalytic converters. In India some automobile companies have plans to fit their automobile engines with catalytic converters.

Nitrogen Oxides, NO_X

It consists of mixed oxides, nitric oxide and nitrogen dioxide (NO and NO_2 respectively)—the former is a colourless and odourless gas but the latter (NO_2) has a reddish brown colour and pungent smell.

The formation of NO and NO_2 is based on the chemical reactions:

$$\underset{\text{(nitrogen)}}{N_2} + \underset{\text{(oxygen)}}{O_2} \xrightleftharpoons{1210° - 1765° C} \underset{\text{(nitric oxide)}}{2NO}$$

$$2NO + O_2 \rightleftharpoons 2NO_2$$

These reactions occur inside the automobile engines so that the exhaust gases consist of NO_X. The latter concentration in rural air is much less than in urban air.

In air NO_X is converted into nitric acid. HNO_3 by natural processes:

$$\underset{\text{(nitrogen dioxide)}}{NO_2} + \underset{\text{(ozone)}}{O_3} \longrightarrow \underset{\text{(nitrogen trioxide)}}{NO_3} + O_2$$

$$NO_3 + NO_2 \longrightarrow \underset{\text{(nitrogen pentoxide)}}{N_2O_5}$$

$$N_2O_5 + \underset{\text{(water vapour)}}{H_2O} \longrightarrow \underset{\text{(nitric acid)}}{2HNO_3}$$

This nitric acid is one of the constituents of acid rain discussed in a subsequent section. From auto exhaust emissions NO_X is removed as discussed above by means of catalytic converters.

Hydrocarbons and Photochemical Smog

Natural processes, particularly trees, emit large quantities of hydrocarbons in air. Methane, CH_4 is a major hydrocarbon. It is generated in large quantities by bacteria formed by anaerobic decomposition of organic matter in water, sediments and soil.

$$\underset{\text{(organic matter)}}{2\{CH_2O\}} \xrightarrow{\text{bacteria}} \underset{\text{(Carbon dioxide)}}{CO_2 \uparrow} + \underset{\text{(methane)}}{CH_4 \uparrow}$$

Domestic animals (cattle, buffaloes, etc.) contribute about 85 million tons of methane to the atmosphere each year. Automobiles are significant sources of hydrocarbons.

In presence of ozone, carbon monoxide, nitrogen oxides and hydrocarbon participate in photochemical reactions (in presence of sunlight). A chain reaction proceeds in which the free radical R $CH_2^{\bullet}$ is generated in the first step. Other free radicals which are formed are: R $CH_2O_2^{\bullet}$ in the second step by reaction with oxygen, R CH_2O^*; R $CH_2O^{\bullet}$ in the third step by reaction with nitric oxide; HO_2 in the fourth step by reaction with oxygen ± a stable aldehyde R CHO is another product at this stage; $HO^{\bullet}$ is formed in fifth step by reaction with nitric oxide (nitrogen dioxide is another product here); and finally, the starting free radical R $CH_2^{\bullet}$ is regenerated by reaction with hydrocarbon, R CH_3 thereby sustaining the chain reaction.

The harmful products in the chain reaction are NO_2 and aldehyde, R CHO. A side reaction also follows by another route through the aldehyde, R CHO; it gives an injurious end product, peroxy acyl nitrate (PAN) which is a strong eye irritant. These reactions lead to photochemical smog formation, which is characterized by brown hazy fumes which irritate the eyes and lungs and also cause serious damage to plants.

Photochemical smog occurs in coastal cities in winter climate, e.g. in Los Angeles, USA which have the heaviest vehicular traffic.

Sulphur Dioxide, SO_2

Sulphur dioxide is a colourless gas with a pungent odour. It is produced from the combustion of any sulphur-bearing material. Sulphur dioxide, SO_2 is always associated with a little of sulphur trioxide, $SO_3^{\bullet}$.

$$\underset{\text{(sulphur)}}{S} + \underset{\text{(oxygen)}}{O_2} \rightleftharpoons \underset{\text{(sulphur dioxide)}}{SO_2}$$

$$2SO_2 + O_2 \rightleftharpoons \underset{\text{(sulphur trioxide)}}{2SO_3}$$

Man-made sources—coal-fired power stations and other industries contribute about 33 per cent of SO_X pollution while natural sources, viz. volcanoes provide about 67 per cent of SO_X pollution.

Soot particles containing metal oxides, catalyze the oxidation of sulphur dioxide to trioxide.

$$SO_2 + O_3 \longrightarrow SO_3 + O_2\ H_2O \quad H_2SO_4 \longrightarrow H_2SO_4)_n$$

(sulphur dioxide) (ozone) (sulphur trioxide) (sulphuric acid) (aerosol droplet)

$$SO_2 + \tfrac{1}{2}O_2 \xrightarrow{\text{soot}} SO_3 \longrightarrow H_2SO_4 \longrightarrow H_2SO_4)_n$$

(aerosol droplet)

The first reaction above occurs in presence of ozone and water vapour. The product, sulphuric acid is formed on aerosol (fine particle suspended in air as in smoke, fog, mist, etc.) droplet. Sulphuric acid is one of the constituents of acid rain.

In winter climate sulphur oxides from thermal power plants along with other gases leads to *smog* formation, e.g. *London smog*. This is known as reducing smog in contrast with photochemical smog which is known as oxidising smog (consisting of hydrocarbons, nitrogen oxides and ozone). *London smog* (1952) is well-known for its disastrous effect. Heavy smog (SO_2) conditions prevailed in London for five days which killed about 4,000 people. The causes of death were bronchitis, pneumonia, and other respiratory troubles particularly among aged people.

Control of SO_X pollution

SO_X (sulphur oxides) from flue gases of industrial plants can be removed by means of chemical scrubbers. The flue stack gases are led through a bed of (slurry) of limestone, $CaCO_3$ (calcium carbonate) which absorbs sulphur dioxide quite efficiently.

The method is economical but the disposal of solid waste, calcium sulphate is a problem.

Alternatively, sulphur oxide in aqueous solution is treated with citric acid salt and the resulting solution is exposed to a stream of hydrogen sulphide gas whereby sulphur is deposited. This sulphur can then be recovered and utilised.

Thermal power plants, major sources of man-made SO_X pollution, are normally constructed with tall chimneys to disperse the emissions over a wide area. This reduces the local problem but creates problems for far away areas through acid rains (*see* below).

Acid Rain

It has been described above that much of nitrogen oxides, NO_X and sulphur oxides, SO_X entering the atmosphere are transformed into nitric acid (HNO_3) and sulphuric acid (H_2SO_4) respectively. These combine with hydrogen chloride, HCl from HCl emissions both by man-made and natural sources) and generate acidic precipitation, known as *acid rain*.

Fig. 6.3. Acid rain in Greece and Italy

Acid rain is a major environmental issue as it badly damages the environment. It damages buildings and structural materials of marble, limestones, slate and mortar. These materials become structurally weak as calcium carbonate reacts with sulphuric acid to form soluble sulphate, which is leached out by rain water:

$$\underset{\text{(Limestone)}}{CaCO_3} + \underset{\text{(Sulphuric acid)}}{H_2SO_4} \longrightarrow \underset{\text{(Calcium Sulphate)}}{CaSO_4} + CO_2 + H_2O$$

In Greece and Italy invaluable stones and statues have been partly dissolved by acid rain. Besides these, acid rain damage forests in Germany and lakes in Sweden and Canada. Acid rain originated from U.K. but far away in Sweden, it damaged some 8,000 lakes of which 4,000 are dead. Similarly, acid rain from USA damaged lakes and forests in Canada. In India, Taj Mahal is threatened by acid rain from Mathura refinery and other industries.

Control of Acid Rain

Acid rain can be checked if its constituents, sulphur oxide and nitrogen oxide are controlled as discussed above.

Particulate

Small solid particles and liquid droplets are collectively termed *particulates*. They originate both from natural and man-made sources. Every year natural sources discharge 800-2,000 million tons and man-made sources 200-500 million tons of particulates. Among man-made sources, fly ash from thermal power plants

Table 6.3: World-wide addition of Particulate matter to the Atmosphere (in million tons)

Particulate matter	Annual Production	
	Natural source	Man-made sources
Total particles	800-2000	200-450
Dust and smoke	—	10-90
Salt, forest fires	450-1100	—
Sulphate	130-200	130-200
Nitrate	30-35	140-700
Hydrocarbons	15-20	75-200

deserve mention. The Table 6.3 gives a list of annual production of particulate matter from the two sources.

Particulates range in size from 0.0002 μ (about the size of a molecule) to 500 μ ($1 \mu = 10^{-6}$ metre). The number of particles in the atmosphere vary from several hundred per cm^3 in clean air to more than 100,000 per cm^3 in highly polluted air (urban/industrial area).

Soot

Soot particles originate from fuel combustion and consist of highly condensed product of polycyclic aromatic hydrocarbon (PAH)—roughly 100 condensed aromatic rings. The hydrogen content of soot is 1-3 per cent and oxygen content 5-10 per cent due to partial surface oxidation. Due to large surface area, soot acts as a carrier for toxic organics, e.g. benzo-α-pyrene and toxic trace metals, e.g. beryllium, cadmium, chromium, manganese, nickel, vanadium, etc.

A soot particle has an average size 0.1-20 μ. The finer particles (< 3μ) are the worst causes of lungs damage due to their ability to penetrate deep in our respiratory tract and thence in lungs where they remain for years and cause all sorts of diseases such an cough, bronchitis, asthma, and finally cancer. Particulates cause increased corrosion of metals which assume serious dimensions in industrial and urban areas. They are responsible for damage to buildings, sculptures, paintings, etc.

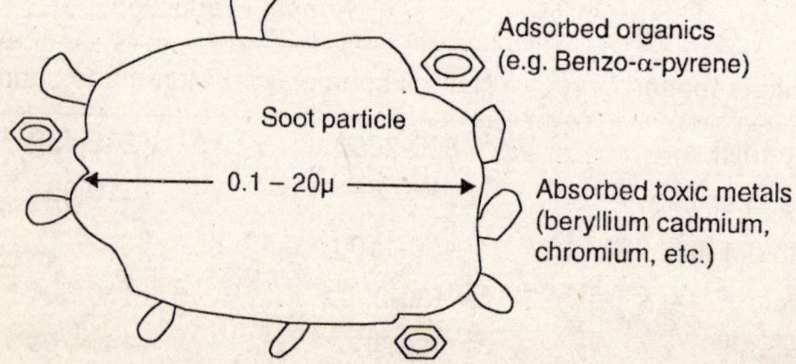

Fig. 6.4. A soot particle

Particulates play key roles in the atmosphere. They reduce visibility by scattering and absorption of solar radiation. They influence the climate through the formation of cloud, rain and snow by acting as nuclei upon which water can condense into raindrops. Atmospheric particulate levels can be correlated with the extent of precipitation over cities and suburbs.

Control of Particulate emissions

The removal of particulate matter from gas streams is an essential step for air pollution control. The best equipment is the *Electrostatic Precipitator.*

Electrostatic precipitator is based on the principle that aerosol particles acquire charge when subjected to an electrostatic field.

$$F = Eq$$

where F = force in dynes to which particles are subjected,

E = voltage gradient and

q = charge on the particles (esu)

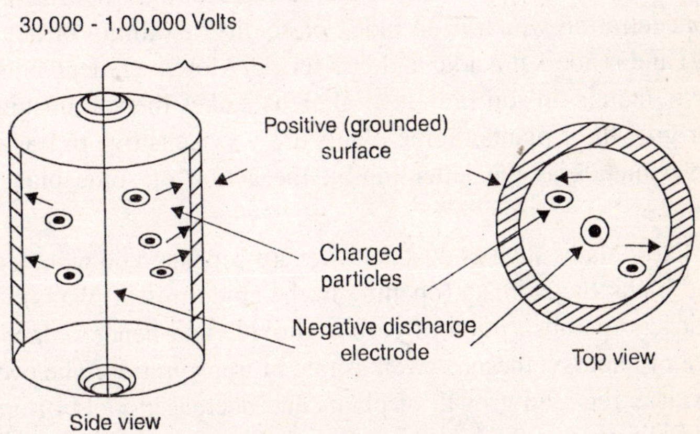

Fig. 6.5. Electrostatic Precipitator

The particles acquire a charge when a gas stream is led through a high voltage (30,000-100,000 volts) DC (direct current) corona (electrical discharge under high voltage). The charged particles are attracted to a grounded surface from which they are recovered.

Air Pollution and Biosphere

Air pollutants are present largely in the troposphere and lower stratosphere. The ground air, 1-100 metres high, is very much polluted in urban and industrial areas. Some pollutants are absorbed on vegetation, buildings and water surfaces.

The primary pollutants discharged into the atmosphere undergo chemical changes in presence of water vapour, oxygen and solar ultra-violet radiation and produce secondary pollutants.

Carbon dioxide $\longrightarrow$ Carbonic acid
Hydrogen sulphide $\longrightarrow$ Sulphuric acid
Sulphur dioxide $\longrightarrow$ Sulphuric acid
Nitrogen oxides $\longrightarrow$ Nitric acid

These pollutants (secondary) have harmful effects on soil, vegetation, crops, animals, men and materials.

Plants are affected both by gaseous pollutants and by particulates deposited on soil. Acid rain over a period of time tends to reduce the soil pH (= log H^+ i.e. negative logarithm of hydrogen ion concentration which is an index of acidity, alkalinity or neutrality) and renders it acidic and less fertile. Moreover, deposition of toxic metals on soil in industrial areas makes the soil unsuitable for growth of plants. Some plants are very sensitive to traces of toxic metals as the latter inhibit the action of some plant enzymes.

Particulates such as dust and soot are deposited on plant leaves and block the stomata (opening in the epidermis of plants). This restricts the absorption of carbon dioxide and hence reduce the rate of photosynthesis as well as rate of transpiration. The overall result is retarded growth of plants and decreased yield of crops. In California, USA the presence of sulphur dioxide in air and metallic pollutants in soil killed vegetation in an area of 300 km^2 and affected growth on a further 350 km^2 of land. In Leeds, U.K. there was drastic decrease in growth of lettuce and radish in heavily polluted industrial areas compared to less polluted areas of the city.

Ozone and peroxyacylnitrate (PAN) (in photochemical smog, see previous section) are oxidizing agents which attack plants by oxidising their sulphydril (-SH) groups of proteins into disulphides. This leads to inhibition of individual enzyme activity. They also affect photosynthesis by plants.

Cattle are affected by air pollution , particularly under smog conditions. They develop breathing troubles, loss of appetite and show low milk yield while many of them die.

Man has become the victim of air pollution. Thousands of chemicals pose the problems of health hazards during manufacture and handling. A typical list of 24 extremely hazardous substances in the atmosphere has been submitted by the United States Environmental Protection Agency (1973):

Acrylonitrile, Arsenic, Asbestos, Benzene, Beryllium, Cadmium, Chlorinated solvents, Chlorofluorocarbons, Chromate, Coke oven emissions, Ethylene oxide, Lead, Mercury, Ozone, Sulphur dioxide, Vinyl chloride, Toxic waste disposal emissions and leachates (washings), etc.

Meteorology and Air Pollution

Air pollution, one of the man-made activities, has some impact on meteorology, i.e. the science of atmospheric phenomenon. Meteorology is based on the physical parameters such as temperature, wind, moisture, and movement of air masses in the atmosphere. It is also affected by the chemical properties of the atmosphere and the chemical reactions going on in the atmosphere.

The air pollutants get dispersed in the atmosphere depending on the patterns of air circulation. In this context *temperature inversion* plays an important role. It occurs when a warm air mass moves above a cold air mass resulting in air stagnation of the latter (cold air) in which air pollutants get trapped. The air above the ground becomes polluted. This happens when warm air blows over a mountain range and over cool air on the other side of the range. Such a phenomenon is observed in Denver, USA of the east of the Rocky mountains.

Human activities are partly responsible for changing the meteorology of the earth. These activities are:

1. Deforestation and loss of forest cover;
2. Shifting of surface water and ground water in large amounts;
3. Release of heat from power plants;
4. Emission of particles and trace gases into the atmosphere;
5. Release of carbon dioxide into the atmosphere by combustion of fossil fuels;
6. Emission from transport system into the lower and upper atmosphere.

The loading of particulate matter into the atmosphere influences the climate. It has been shown in Table 6.3 that each year natural sources are injecting about 800-2,000 million tons while man-made sources about 200-400 million tons of particles in the atmosphere. As already discussed, particles induce cloud formation and rainfall. It has been calculated as an approximation that if the particle loading increases by 50 per cent, the average temperature of the earth will decrease by about 0.5°C to 1°C due to particle-induced cloud formation. This partly counterbalances the temperature rise due to Greenhouse Effect.

6.6. AIR QUALITY STANDARDS

Each pollutant, present in air, has a *threshold limit value (TLV)* which, if exceeded, causes public health hazards. Table 6.4 gives a list of typical pollutants with their threshold limits (TLV). For factory workers TLV sets the limit of exposure for 40-hour week (8 hours a day) without adverse effects. These TLV values are determined mainly by experiments on animals.

6.7. SOIL POLLUTION

Lithosphere with humus cover is known as soil. It is a limited natural resource but most important resource which supports life and contributes to the development of the country. Indeed the soil quality of a nation is its most precious material heritage.

Table 6.4: Threshold Limit Values (TLV) for some common Pollutants (Gases and Vapours)

Pollutant	Threshold Limit Values	
	ppm	mg/m^3
Acetone	750	1780
Ammonia	25	18
Arsenic	0.2-0.5	–
Benzene	10	20
Cadmium	–	0.05
Carbon dioxide	5000	9000
Carbon monoxide	50	50
Carbon tetrachloride	5	30
Chlorine	10	30
Chloroform	10	50
Hydrogen chloride	5	7
Hydrogen sulphide	10	14
Lead	–	0.2
Nitric oxide	25	30
Ozone	0.1	0.2
Sulphur dioxide	2.0	5.0
Vinyl chloride	5	10

The top soil provides water and all the nutrients required by plants for their growth. The survival of man depends upon agricultural crops, grassland and trees which grow on soil and supply food, fibre, timber, fuel and other products. But it faces threat of erosion from man-made activities. It takes 500-1000 years to build up one inch of top soil while loss of soil takes only a few years. Deforestation alone accounts for loss of 6000 tons of top soil each year.

The soil pollution sources are:

(i) indiscriminate solid waste disposal on land;

(ii) random discharge of industrial waste effluents on land and water courses;

(iii) excessive use of pesticides, herbicides and chemical fertilisers in agriculture;

(iv) excessive irrigation;

(v) open defecation on land by animals and man particularly in rural areas and city slums;

(vi) radioactive materials from radioactive fall-out arising nuclear explosion tests, reactor accidents etc.

Dumping of solid wastes on land is quite harmful since heavy metals and toxic elements present in tense wastes contaminate the soil and ultimately seep into underground water after rainfall. It is estimated that about 15 million tons of solid wastes are generated every year by more than 110 million urban citizens in India.

In general, in rural areas poor illiterate people have the nasty practice of open defecation in absence of sanitary toilets. Both human and animal excreta contain many pollutants which degrade not only soil but also nearby ponds and canals where they mix after being washed by rain water.

Many industries generate huge volume of liquid effluents and solid wastes which are mostly thrown on land and in water courses (streams and rivers). These effluents and wastes contain phosphororus, chromium, lead, zinc, nickel, cyanides etc. which affect the quality of soil and render it injurious for plants and animals and ultimately man. Agricultural run-off (wash water from agricultural fields) contains lot pesticides, fertilisers, herbicides which find their way into water courses causing water pollution in addition to soil pollution. Excessive irrigation increases the salt content of soil and makes it saline, thereby decreasing soil fertility.

Each nuclear test is accompanied by radioactive fall-out which spreads radioactive contaminants in nearby areas over land and water bodies. Moreover, radioactive wastes from atomic power plants pose longterm health hazards after they contaminate land and water bodies.

Control of Soil Pollution

Solid waste disposal methods are actually the methods for control of soil pollution. Non-hazardous wastes are managed by two techniques (i) Landfill and (ii) incineration. Miscellaneous waste materials—food wastes, vegetable and animal wastes, paper, wood, rubber, leather, plastics, tin cans, crockery, glass, etc. are deposited in a regulated manner in general pit in backyards of houses and municipal areas. These are then covered with 15 cm. thick soil. The wastes may be burnt from time to time. Another option for municipal wastes is to use biological process: organic wastes are allowed to be decomposed into humus-like substances after removal of iron and metallic particles. In presence of stream of air, aerobic micro-organisms ensure rapid decomposition of the wastes. Such compost plans are in operation in several parts of the country e.g. Ahmedabad, Mumbai, Chennai, Delhi, Pune, Hyderabad and Kolkata.

The second method viz. Incineration is the preferred technique for waste management particularly in developed countries. It reduces the bulk of waste by 90 per cent at 900-1000°C. This is an environment-friendly technique — free from corrosion, emission of offensive odours and from bacteria. The waste heat can be utilised for supplementing electricity generation for domestic heating. However, the technique is costly at present requiring expensive equipment.

Very recently scientists are developing solid waste management by Biotechnology using micro-organisms and also genetically engineered micro-organisms.

6.8. RADIATION POLLUTION

Radioactive substances such as Uranium, Thorium, Plutonium, Strontium, Radium, Cadmium, Tritium, Carbon-14, Iodine-131 etc. emit radiation which is harmful to living organisms, cause genetic damage to man and death from cancer. This is the phenomenon of *Radiation Pollution*, which has long-term effect on environment. Nuclear reactors and nuclear material processing plants are fixed sources of radiation pollution.

Most dangerous sources are nuclear tests and fall-out from nuclear bomb explosion which spreads over wide area and the radioactive materials/ash are detrimental among all the environmental segments — air, water, land and biosphere.

Background Radiation

Cosmic rays from upper atmosphere and also radionuclides in earth's crust (e.g. uranium-238, uranium-235, thorium-232, radium-224, radon-222, potassium-40 and carbon-14) are sources of background radiation reaching the biosphere. Though usually the background radiation is below tolerable level, the level may be exceeded and it may become critical on exposure to man-made radiation (mining and refining of radioactive minerals, nuclear power plants and reactors, nuclear test explosions etc.)

Effects of Radiation Pollution

Radiation pollution is more fatal than any other kind of pollution. Man-made radiation has increased in quantity since the invention of atom bomb and use of nuclear energy for reactors and power plants. Radiation damages living cells by ionisation leading to skin burn, cancer and anaemia which are beyond medical treatment. It may be noted that the discoverer of radiation were themselves the first victims — W.H. Roentgen who discovered X-rays died of bone cancer in 1923 while Madam Curie, the discoverer of radium died of anaemia and cancer. They were not aware of the precautions against radiation.

The horrors of Hiroshima and Nagasaki (August 6, 1945) in Japan have not been forgotten. The two atom bombs instantly killed about 6 lakhs people, destroyed the two cities and unleashed radiation which has caused generations to suffer from various diseases including genetic disorder. Radiation still continues to damage plants, soil and biosphere in the region even after 55 years.

The worst nuclear accident occurred at *Chernobyl*, USSR (now

CIS) on *28th April, 1986*. The reactor exploded as a result of runaway nuclear reactions — radioactive fuel and debris shot up in the air like volcanic explosion and spread far out in the surrounding areas. The accident killed about 2000 people and damaged soils, water and vegetation in an area of 60 sq. km. around Chernobyl. Several generations in the region will suffer from genetic damage.

For control of radiation pollution it is necessary to take utmost precautions regarding safety devices for reactors and strictly ban and enforce the ban on nuclear test explosions. Leakage from reactors, careless handling and use of radioactive fuels, fission products and radioisotopes must be stopped. The practice of dumping radioactive wastes into seas and oceans should also stop. These wastes should be buried underground for a long period to ensure that the radioisotopes loss their activity.

6.9. NOISE POLLUTION AND HEALTH

Noise is part of our environment. With progress in industrialisation, the noise level has been rising continuously. In the 19th century the development of the steam engines, petrol engine and machines in factories resulted in increasingly noisy environment. In the 20th century this was further accelerated by introduction of diesel engine, jet engines, turboprop, high tech machineries, construction site machineries and automobile traffic. Noise has been recognised as one of the dimensions of pollution which brings about degradation of the environment and creates health and communication hazards.

Sound and Human Acoustics

Sound consists of wave motion in an elastic medium such as air, water or solids (e.g. metals, plastics, wood, bricks, etc.). Sound waves travel through the medium from the source to the recipient or listener. The rate of the oscillation of the medium is known as the *frequency* of the sound, the unit being *Hertz (Hz)* or *cycles per second*. The frequency is a measure of the pitch of the sound received by the listener. High frequencies mean high pitched sounds which are more irritating to the individual than low frequencies. The second parameter of sound in *sound pressure* which is measured in Newtons per sq. metre (N/m^2). The third parameter on sound is its *intensity*, expressed in watts per sq. metre, i.e. the quantum of sound energy that flows through unit area of the medium in unit time.

The human ear receives sound waves which set up oscillations in the ear drum (tympanic membrane). These oscillations cause movement of three small bones in the middle ear behind the ear drum. These then pass through the fluid in the inner ear to the auditory nerve and finally transmitted to the brain. The oscillations or sound are intensified and interpreted in the brain, which can select sounds into different categories — speech, music, noises, etc.

The sensitivity of the ear varies from person to person. With ageing, people lose hearing power gradually. A young person, 18 years old, with normal hearing, has audio range between 20 Hz and 20,000 Hz. The audio sense is sharpest in the frequency rage 2000-8500 Hz.

Noise Measurement Units

As mentioned before, sound pressure and sound intensity are two important parameters of noise. The common scientific acoustic unit is the Decibel (dB). It is not an absolute physical unit like volt, metre, etc. but is a ratio, expressed in logarithmic scale relative to a reference sound pressure level.

$$1 \text{ decibel (dB)} = 10\log_{10}\frac{\text{intensity measured}}{\text{reference intensity}}$$

The reference intensity used in the threshold of hearing which means sound which can be first heard at a sound pressure of 2×10^{-5} Newtons per sq. metre or sound intensity of 10^{-12} watts per sq. metre.

Noise meters have been designed for noise measurement from

low to high frequencies, characteristic of human ear capacity. These meters record the dB scale for routine measurement of general noise levels. Refined noise meters have been developed to take care of peak noise levels, duration of noise exposure and quality of noise which are aspects of specified noise situation.

Table 6.5: Sound Measurement (Intensities, Pressures and Decibels) in air at room temperature and sea level pressure

Intensity (Wm^{2-})	Pressure (Nm^{2-})	dB	Sound Source
100	2,00,000	200	Saturn rocket take off
1.0	20	120	Boiler shop
10^{-2}	2.0	100	Siren at 5 metres
10^{-4}	0.2	80	Heavy machinery
10^{-6}	0.02	60	Normal conversation at 1 metre
10^{-8}	0.002	40	Public library
10^{-12}	2×10^{-8}	0	Threshold of hearing

L_{10} (18 hour) Index

This is used for road traffic measurement, adopted in UK for noise legislation. The index is expressed in dB—it is the arithmetic average hourly values of the noise level exceeded for 10 per cent of the time over 18 hours between 6:00 and 24:00 hours on any normal week day. It includes peak noise values and fluctuation of noise depending in the type of vehicle and traffic density.

Effective Perceived Noise Level (L epn)

This is recommended for aircraft by the International Civil Aviation Organisation (ICAO) as the standard for use in noise evaluation. The index is based on the scale equivalent to the dB scale +13 and takes care of the peak frequency of jet aircraft noise as well as duration of aircraft flyover.

Noise Classification

There are broadly three categories of noise:

(i) Transport noise,

(ii) Occupational noise, and

(iii) Neighbourhood noise.

Transport Noise

Transport noise can be further sub-divided into (i) Road traffic noise, (ii) Aircraft and (iii) Rail traffic noise.

Road traffic noise

Traffic noise is increasing over the years with increase in the number of road vehicles. Traffic speed is the major cause of noise. The noise volume is enhanced with increase in traffic speed. Modern highways and traffic system encourage higher speeds.

In general, on urban roads there are distinct traffic peaks in the morning and evening (10 A.M. and 6 P.M.) as people travel to and fro workplaces. Heavy diesel-engined trucks are the noisiest vehicles on roads at present. The permissible noise levels for cities are prescribed by the Central Pollution Control Board, India:

Areas	Day	Night
Industrial	75 dB	65 dB
Commercial	65 dB	55 dB
Residential	50 dB	45 dB
Sensitive areas up to 100 m around hospitals, schools	50 dB	40 dB

These limits are, however, violated in all big cities in India, Calcutta being the worst case. The average noise levels in busy streets in Calcutta during rush hours (between 10:30-12:00 hrs. and 18:00-19:30 hrs.) are 90 dB. People live in an environment of noise generated by blasting horns, rumbling tyres and screeching brakes. Awful road accidents contribute to the misery.

Aircraft Noise

The noise levels have peak values when aircrafts fly low and overhead or take off and land at airports. The noise limits prescribed by UK airports for take-offs are 110 PNdB (1 PN dB = dB scale + 13) during day and 102 PNdB during night. These may be compared with the values in USA: 112 PNdB during day at New York.

Rail traffic

It is less of a nuisance as compared to the previous types of traffic noise.

Occupational Noise

Industrial workers are exposed to noisy working environment for 48 hours a week (8 hrs. a day for 6 days a week). Some typical occupational noise levels are given below:

Table 6.6

Industrial Source	Noise Level (dB)
Steel plate riveting	130
Oxygen torch	126
Boilers' shop	120
Textile loom	112
Circular saw	110
Farm tractor	103
Newspaper press	101
Bench lathe	95
High speed drill	85
Supermarket	60

Millions of workers suffer from progressive hearing damage and become prone to accidents under their working conditions. Their working efficiency is also affected.

Neighbourhood Noise

Loud TV and radio sets, loud cassettes, loudspeakers in public functions, disco music, etc. are sources of neighbourhood noise which disturb and irritate the general public and also harm the patients.

Noise Pollution Hazards

The human ear drum is struck by noise in the form of air-borne mechanical energy. While the tolerable conversation level is 65 dB at a distance of 1 metre, 125 dB gives the sensation of pain in the ear and 150 dB might be a killer.

High intensity noise for continuous periods is the major cause for ear damage. If a noise level exceeding 90 dB in the mid-frequency range reaches the ear for more than a few minutes, then the sensitivity of the ear is reduced.

Noise pollution can cause pathological or psychological disorders. High frequencies or ultrasonic sound above the audible range can affect the semi-circular canals of the inner ear and make one suffer from nausea and dizziness. Mid-audible frequencies can lead to resonance in the skull and thereby affect the brain and nervous system. Moderate vibration can also cause pain, numbness and cyanosis (blue colouration) of fingers while severe vibration results damage to bones and joints in the bands with swelling and stiffness.

In industrial and other establishments the general impact of noise pollution is lower efficiency, reduced work rate and higher potential for accidents and injuries.

In residential areas even low frequency noise of 50-60 dB at night disturbs sleep, particularly among the aged people, causing adverse effect on health.

Children, exposed to excessive noise, show signs of behavioural disorder which in later age develop into destructive nature and neurotic disorders in the adult.

Excessive noise is one of the major factors for chronic exhaustion and tension in our daily lives. This may explain why more and more people tend to become addicted to alcohol, tobacco and drugs.

Noise pollution has also impact on travel of migratory birds from winter to tropical climate. Thus the increase of noise pollution in Calcutta and construction of high-rise buildings near Alipur Zoological garden have led to decline in the number migratory birds from CIS (former USSR) from 15,000 in 1980 to 2000 in 1990.

Permissible Noise Levels

In this age many people work and live in environment where the noise level is not hazardous. But over the years they suffer from progressive hearing loss and psychological hazards. The maximum permissible noise levels are summarised as follows.

Table 6.7: Maximum permissible Noise levels

Situation	Permissible Noise, dB
Road traffic near residential areas	70
Ear protection required	85
Factory work (48 hr. week)	105
Prolonged noise causing permanent damage	100
Threshold of pain (30 sec. Duration)	120
Maximum for sonic boom	150
Ear drum rupture	180
Lungs damage	195

6.10. TOXICITY IN FOOD CHAIN

Toxic elements (those which injure organisms) present in air, water and soil pass from one organism to another through food intake (*Food Chain*). Thus polluted air, water and soil pass on the toxic substance to green plants (trophic level), then to herbivores i.e. primary consumers who feed on plants and leaves which then pass on to the carnivore or secondary consumers and finally to man, the omnivore who eats both plants and animals. Presence of toxic elements in terrestrial and aquatic food chain causes health hazards for man.

Toxic chemicals in air include arsenic, asbestos, benzene, beryllium, cadmium, chlorinated solvents, lead, mercury, ozone, sulphur dioxide, etc. while those in water arsenic, cadmium, beryllium, chromium, copper, fluoride, lead, manganese, mercury, molybdenum, zinc, etc. and also pesticides. The water bodies contain a large number of pesticides, mainly from darning of agricultural land by rainwater (agricultural run-off). Some of these pesticides such as DDT (dichloro diphenyl trichloroethane) is a persistent chemical in the environment. Once introduced, it keeps on circulating for many years. It is important to note that DDT concentration keeps on building up and magnifying at successive trophic levels: plankton-clams-fish-birds-man. The process of magnification or amplification at each trophic level in the food chain is called *biomagnification*. Thus 0.04 ppm DDT in plankton ends up in fish at 0.2 ppm and birds at 75 ppm. It is known that even penguins at faraway Antarctica pick up DDT from marine fish. Those with high level of DDT produce eggs which are too thin and fragile and are threatened with extinction.

QUESTIONS (Unit 5 and 6)

1. "Carbon dioxide is a non-pollutant but it is of much environmental concern" — Elucidate with reference to the Greenhouse Effect.
2. What are Greenhouse gases? What are their relative contribution to the Greenhouse Effect?
3. What is Global Warming? Discuss its impact on global climate, food production and topography.
4. Name the primary air pollutants and their respective sources.
5. How would you control sulphur dioxide and particulate emissions from thermal power plants?
6. Write notes on:
   (a) Ozone holes
   (b) Acid rain
   (c) El Nino
   (d) Photochemical smog

7. What is meant by Threshold Limit Value (TLV) of a pollutant? State TLV for (a) acetone, (b) ammonia, (c) benzene, and (d) carbon monoxide.
8. Discuss the role of Forests in the Ecosystem.
9. What are the causes of Deforestation?
10. Discuss the effect of "Jhum Cultivation" on N.E. India's environment.
11. What are the adverse effects of Desertification? Suggest remedial measures for Desertification.
12. What are the sources of fresh water? Give their quantities in terms of total percentage of water resources.
13. List the uses of water by man giving their relative percentages.
14. Give an account of domestic waste water treatment.
15. Give a flowsheet diagram of municipal waste water treatment (primary and secondary)
16. Write notes on:
    (a) Ganga Action Plan
    (b) Desalination of sea water.
17. What is the importance of soil in national economy?
18. List the factors contributing to land degradation.
19. How can soil erosion be checked?
20. Suggest methods for control of soil pollution.
21. What causes Radiation Pollution?
22. How does Radiation Pollution lead to health hazards?
23. Discuss the harmful effects of Radioactive fallout from a nuclear test explosion.
24. What is Food Chain?
25. How does a Food Chain get contaminated? How is it possible to eliminate such contamination?
26. How does DDT accumulate in Food Chain? Explain Biomagnification.

For Objective Tests *see Unit 11*.

Unit 7

Effects of Environment on Biosphere

7.1. EXTINCTION OF SPECIES

The earth was born about 4.5 billion years ago. The first life form appeared in the oceans about 3.5 billion years ago and man came last of all some 20 million years ago. Thus in the process of evolution of life man appeared only when the proper stage was set for optimum conditions of environment for all life forms to exist. In spite of the relatively recent origin of man in the geological time scale, he wields immense authority on the biosphere.

Within the biosphere exist several million different kinds of organisms sufficiently different from one another, called '*species*'. These species are independent units of life but they have interactions with each other to form a community settled in a habitat under the same ecosystem. These interactions are important for their proper functioning.

There are some 5-10 million species on earth which are the products of 3 billion years of evolution involving mutation, recombination and natural selection. But changing environments like ice age followed by warm periods and even drought exerted severe selection pressures ultimately leading to extinction of species and evolution of new species. The present day life probably constitutes about 1 per cent of the total species that has existed on earth so far. Natural extinction has gone side by side with man-made extinction. However, the present wave of extinction is essentially man-made which has far exceeded natural extinction due to ever-increasing needs and greed of man. By the end of the last century about 1 million species have vanished. This is an ecological crisis which has threatened the whole life-support system and the large number of habitats and species particularly in the developing countries. This is called *specide*, which is a worse crime than genocide (Swaminathan, 1978). According to a report of the World Commission on Environment and Development (1987) the present man-made extinction is hundred times higher than natural extinction.

Human activities responsible for species vulnerability or extinction are: industrialisation, urbanisation, intensive agriculture, large scale deforestation, destruction of natural habitats of species, exploitation of natural resources, overgrazing of grassland and forest land, poaching and hunting, trade in timber and wild life, nuclear test explosions etc.

7.2. ENDANGERED AND THREATENED SPECIES

According to the International Union for Conservation of Nature and Natural Resources (TUCN) the "*endangered species*" is defined as "the categories of plants and animals whose number and quantity have been reduced to critical level and are in immediate danger of extinction." TUCN has also defined **vulnerable species** as "species which are rare and under threat or actually decreasing in number or species seriously depleted in the past and not yet recovered" and **extinct species**" as those that cannot be found in areas they once inhabited or in other likely habitats".

In 1994 the minimum estimate for numbers of globally threatened animal and plant species was about 5400 animals and 26,000 plants. About 11 per cent of bird species, 18 per cent of mammals, 5 per cent of fish and 11 per cent of plants were catalogued as threatened. The Red Data Book of IUCN shows more than 1000 creatures are threatened with extinction, some very soon and some within 100 years. Animals facing immediate danger of extinction are rhinoceros, particularly of the Indian variety, the Royal Bengal Tiger, Siberian Tiger, Mexical grizly bear, red wolf, mountain gorilla, Asiatic lion, Arabian orys and Japanese golden butterfly.

A list of endangered Indian animals, prepared by CITES (Convention of International Trade in Endangered Species of Wild Flora and Fauna), includes:

Mammals: Golden langur, Gibbons, Gangetic dolphin (sisu), Great Blue Whale etc.

Carnivore: Wolf, Asian black bear, cheetals, golden cat, Asian lion, Royal Bengal Tiger etc.

Herbivore: Rhinoceros, Musk deer, Indian bison, yak etc.

Birds: Pink-headed ducks, Imperial eagle, Great white crane, Indian bustard, Bengal florican, Pied hornbill, etc.

Reptile: Crocodile, Gharial, Turtle

India is the home of about 200,000 species. Among these 1500 plant species, 80 mammals, 45 birds, 15 reptiles and 3 amphibians belong to the endangered species.

7.3. BIODIVERSITY AND BIORESERVE

The **Earth Summit at Rio de Janeiro, Brazil (1992)** highlighted the importance of Biodiversity by urging all the countries in the world for international co-operation for conservation of all living species. Biodiversity has been defined (Convention on Biological Diversity, Nairobi, 1992) as "the variability among living organisms from all sources including terrestrial, marine and other aquatic ecosystems and the ecological complexes of which they are part — this includes diversity within species, between species and ecosystem." It implies two related concepts: *Genetic diversity* and Ecological diversity. Genetic diversity is the amount of genetic variability among individuals of a single species as also between species. *Ecological diversity* is the number of species in a community of organisms. It is essential to maintain both kinds of diversity for the functioning of ecosystems and hence for human welfare.

The Indian region (8° - 30°N and 60° - 97°E) with a total area of 329 million hectares is very rich in biodiversity. It is estimated that over 45000 species of plants (flora) and 75,000 species of animals (fauna) occur in this country. The vascular flora which forms the conspicuous vegetal cover itself comprises 15,000 species of which more than 60 per cent are endemic and do not occur elsewhere. Besides flowering plants, there are a large number of non-flowering plants such as ferns, liverworts, algae and fungi.

It should be realised that biodiversity holds the key to human civilisation. Loss of biodiversity due to extinction of plant and animal species is a severe blow to the earth's ecosystem which ultimately brings disaster for man, threatening his survival on earth. Our food requirement is met by various plants, animals, birds and fish. We get our drugs and medicines from many medicinal plants. Can we imagine a situation if *Penicillium* had been extinct before man could use it as an antibiotic or if *Cinchona* became extinct before quinine was discovered as a cure for malaria? It is, therefore, in our own interest to conserve our plant as well as animal and micro-organism wealth.

Biosphere Reserve

It has been recognised throughout the world that the conservation of biodiversity should be given top priority in any national agenda. Ancient India set the tradition of caring, conserving and worshipping trees and animals. The late Prime Minister Indira Gandhi emphasised while launching the World Conservation strategy in India in 1980: "The interest in conservation is not sentimental but the rediscovery of a truth well-known to our sages. The Indian tradition teaches us that all forms of life—human, plants and animals—are so inter-woven that disturbance in one gives rise to imbalance in other."

UNESCO launched a world-wide project in 1971 for conservation of biotic community viz. *Man and Biosphere Programme* (MAB). This gave rise to *Biosphere Reserve Programme* which was implemented in most of the countries in 1985. The objectives of the programme are:

(i) conserve representative samples of ecosystems;
(ii) provide long-term *in-situ* conservation;
(iii) promote and facilitate basic and applied research and monitoring;
(iv) provide opportunities for education and training;
(v) promote appropriate sustainable managements of the resources;
(vi) disseminate the experience so as to promote sustainable development elsewhere; and
(vii) promote international co-operation.

Bioreserve is an area of terrestrial or coastal environment representing one of the world's natural or biogeographical regions.

It covers as many kinds of plants and animals as possible and includes a great variety of landscapes.

The National MAB Committee has identified 12 sites as potential areas as Biosphere Reserves:

Biosphere Reserve	State
Nilgiris	Tamil Nadu, Karnataka and Kerala
Namdapha	Arunachal Pradesh
Nanda Devi	Uttar Pradesh
Uttarakhand	Uttar Pradesh
North Islands of Andamans	Andaman & Nicobar
Gulf of Mannar	Tamil Nadu
Kaziranga	Assam
Sunderban	West Bengal
Thar desert	Rajasthan
Mannas	Assam
Kanha	Madhya Pradesh
Nokrek (Tura range)	Meghalaya

Biodiversity conservation is based broadly on *In-situ Conservation* and *Ex-situ Conservation.*

In-Situ Conservation: This is the ideal system of conservation through thc maintenance of genetic resources within their habitats i.e., natural or man-made ecosystem in which they occur. A community in harmony with a stable environment (the stability being subject to the general vagaries of natural environment) is the ideal model of long-term conservation. Such conservation includes a network of National Parks, Sanctuaries, Nature Reserves, Natural Monuments, Cultural Landscapes, Biosphere Reserves, etc. which serve as treasure houses for biodiversity.

Ex-Situ Conservation: This is conservation outside their habitats by perpetuating sample populations in genetic resource centres, zoological gardens, botanical gardens; in germ plasm banks for seeds, pollen, semen, ova, cells, etc. Seed banks are good examples of ex-situ conservation where seeds are stored (cold storage) for long durations. Botanical gardens play important roles in the conservation of plant species (rare, endemic and threatened plants).

Sanctuaries: For wild life management in India there are 73 National Parks and 416 wild life sanctuaries. The first National Park in India was the Corbett National Park during British period. A **Sanctuary** is generally a protected area where killing or capturing any bird or animal is prohibited by law.

The following Table gives a list of National Parks and Wild Life Sanctuaries in North East India:

	States	No. of National Parks	No. of Sanctuaries
1.	Arunachal Pradesh	2	4
2.	Assam	2	7
3.	Manipur	2	–
4.	Meghalaya	2	3
5.	Mizoram	–	1
6.	Nagaland	–	3
7.	Tripura	–	4

Kaziranga in Assam is famous for the preservation of one-horn rhino variety. There is proposal to elevate it to a biosphere reserve. Namidhapa sanctuary in Arunachal Pradesh is under consideration for recognition as a bioreserve. Balpakram in Garo hills (Meghalaya) has a wild life sanctuary.

In collaboration with IUCN (International Union for Conservation of Natural Resources) the Government of India operates some special projects for the preservation of endangered species.

(1) *Project Tiger*: There are 18 Tiger Reserves in 13 states — notably the Sunderban Tiger Project in W. Bengal, Namidhapa Tiger Reserve in Arunachal Pradesh and Manas Tiger Project in Assam.

(2) *Gir Lion Project*: This is in the Gir forest in Saurastra, Gujarat which is the only habitat of Asiatic lion.

(3) *Rhino Conservation Scheme*: This is at Kaziranga sanctuary in Assam for management of rhino habitats.

(4) *Crocodile Breeding Project*: This Project is conducted in Andhra Pradesh and Orissa for increasing the crocodile population in fresh water and salt water.

7.4. GENETIC RESOURCES

The world of plants, animals and microbes comprise genetic resources which are identified on the basis of genes. These genes are units or tiny particles of chromosomes found inside the nucleus of reproductive cells. They are actually a small portion of the molecule of Deoxyribonucleic acid (DNA) which carry hereditary instruction or are the bearers of heredity. They determine the shape, appearance and behaviour of a particular species of organism. The genes explain why all numbers of the same species looks alike. Variations between the members of the same species are caused by environmental stress which increases or reduces the role of genes to carry on their natural functions.

Recent advances in the science of Genetics opened up new era of study: *Genetic Engineering* which has immense potential. It deals with the study and analysis of chromosomes, genes and DNA structure and seeks to apply the knowledge in selective breeding of crops, plants and animals for the welfare of mankind.

Genetic resources are the most precious but very vulnerable

and irreplaceable asset in the natural resources available to mankind. They are nature's tools for harvesting solar energy and processing mineral resources into unable food, fibre, timber, fuel, medicines, useful industrial chemicals etc. Thus they comprise all plants, animals and micro-organisms of major socio-economic value for human survival in the physical environment. They are the product of millions of years of evolution and being biologically active entities, they are extremely labile and once they are lost, they cannot be replaced at any cost.

The main reason for remarkable biodiversity in a single country like India is the great diversity of ecosystems which it has supported down the ages. Almost every major type of habitats found here. The position of the Indian sub-continent at the confluence of three biogeographic realms is an important contributory factor and explains the presence of African, Mediterranean, European, Sino-Japanese and Indo-Malaysian elements in the flora and fauna in India. This remarkable diversity has made India a valuable "*gene bank*" for a number of food crops, forests, trees, medicinal and aromatic plants and domesticated animals. There is, therefore, an urgent need to arrest species extinction and use genetic engineering for conservation of biodiversity.

QUESTIONS

1. What is meant by extinction of species?
2. State the causes for species extinction. How can they be checked?
3. What do you mean by endangered species? Give examples.
4. What is Biodiversity?
5. Justify the statement: India is rich in Biodiversity.
6. Discuss the concept of Biosphere Reserve.
7. List the Biosphere Reserves in India.
8. What is genetic resource? Discuss its importance.
9. Explain the concept of "gene bank".
10. List the number of National Parks and Wild Life Sanctuaries in North East India.
11. Distinguish between "In-situ Conservation" and "Ex-situ Conservation".

Unit **8**

Environment as Resource

8.1. CONCEPT OF RESOURCE

Resource is defined as a form of matter and energy which is useful and essential for the functioning of organisms, populations and ecosystems. For human beings, a resource is any form of matter or energy, utilised for satisfying the needs—physiological, economic, social and cultural—both at the individual and community levels. Thus resources include food, air, water, land, soil, forest, fish, solar energy, coal, petroleum, mineral oils, plants, animals etc.

Man depends in nature for his daily needs and sustenance. For clothes we depend on cotton, for furniture we depend on wood i.e. trees in forest and for household goods we depend on machines, tools made of metals extracted from mineral ores. Thus forest, cotton, soil and minerals are our *resources*. Besides for our daily living we need air for breathing, water for drinking etc. so that these are also resources. Other resources include vegetables (agricultural crops), milk (from cattle), fish, animals, energy (fossil fuel, petroleum, solar power etc.) etc.

Besides natural resources as stated above, a significant resource is the human resource. Only through skill and technological knowledge of man, natural resources can be properly and meaningfully utilised. Human resource development has recently been given top priority in national policy. Education and technical skill are essential for development of agricultural, mineral and other resources which contributed to the Gross National Product (GNP) of a country. The United Nations Development programme gives equal emphasis on GNP and Human Resource Development.

8.2. TYPES OF RESOURCES—RENEWABLE AND NON-RENEWABLE

Resources are classified as Renewable and Non-renewable.

Renewable resources are those which get exhausted within a few years but can be regenerated and used again. Examples are plants, forest, water, solar energy etc. Forests are renewable resources since after felling of trees for wood, fuel etc., they can be restored by plantation of new trees (afforestation). Solar energy from the sun is a continuous source of energy which will last till the life time of the Sun (several billion years).

Non-renewable resources are those which, after being exhausted, cannot be renewed. For example, fossil fuel—coal, petroleum, natural gas and minerals—took thousands of years for their formation under the earth's surface. They get exhausted due to over-exploitation and thereafter they cannot be restored or renewed.

8.3. OVER-CONSUMPTION AND CRISIS

In this section we shall deal with some of the principal resources and how their excessive use or consumption as at present will lad crisis due to their shortage.

(i) **Land Resource:** Land is used for many purposes—construction of houses and buildings, roads, railway lines, airports, agriculture, grazing of cattle, setting up of industries etc.

In India the total available land is about 300 million hectares. It is, however, a fixed asset. As a result of our population explosion, we possess the lowest man: land ratio, barely 0.50 hectares per capita due to use and abuse over the centuries.

The percentage break-up of land use pattern in our country is:

(i) Waste land, 24%

(ii) Pasture land, 4%

(iii) Forest land, 21% and

(iv) Agricultural land, 51%

Considering the size of India, available pasture land is insignificant while forest land has been reduced due to conversion of forests into agricultural land and also for housing, roads, industries etc.

(ii) **Soil:** We depend on soil for our daily food and also for meeting our needs for fibre, timber, firewood etc. Hence soil is indeed an important resource. But this resource is getting depleted mostly due to erosion and degradation. Large scale deforestation and intensive grazing of grassland have exposed the soil to the weathering forces—water, wind erosion. Apart from these, the prevailing practice of jhum cultivation in the N.E.region, canal irrigation, road construction and surface mining also favour soil erosion. Top soil (most fertile zone) is lost as a result of soil erosion leading to loss of productivity of land. It is estimated that every year 25 billion tons of top soil are lost from farm land by erosion. In India about 80 million hectares of cultivated land is subject to severe erosion and about 30 per cent of the total eroded soil is permanently lost to the sea. This is a critical issue as it affects our survival in terms of supply of food, fodder, fuel and daily necessities.

(iii) **Forest Resources:** Forests are valuable renewable resource. They are closely related to vegetation, grassland and wild life resources. They have contributed substantially to the economic development of the country by providing goods and services to the people and industries. Furthermore, they play a very major and vital role in enhancing the quality of the environment by influencing the life support system. Forests are also linked with our culture. As a matter of fact, Indian philosophy and religion grew up by our *rishis* in the forests in total communion with nature.

India has 64 million hectares of forest land out of a total geographic area of 329 million hectares. It is on record that the country had 80 per cent forest cover 5000 years ago which has been reduced barely to 21 per cent. Per capita forest land is 0.1 hectare as compared to the world average of 1 hectare. This shows the poorest record of India. We are losing 1.3 million hectares of forest cover every year and at this rate of deforestation we are likely to lose all the forest and face disaster within the next 25 years.

During 1951-2000 i.e. 50 years after Independence, India witnessed maximum loss of forests, namely, 6.0 million hectares of forest area. This overconsumption has been due to population explosion (both human and livestock population) accompanied by rapid urbanization and industrialisation as well as change of life style. There has been tremendous pressure on forest systems for ever-increasing demands on fuel or fire wood, timber and pulp wood for housing, furniture, agricultural implements, railway slippers, sports goods, ship and boat building, paper and pulp industries etc. Annual consumption of fire wood in the country is about 170 million tons and 10-15 hectares of forest cover is destroyed each year to meet fuel requirements. The consumption of fire wood is 80 per cent in villages and 20 per cent in urban areas. There is a shortfall of about 195 million m^3 in the demand and supply of fuel wood. Similarly the gap for timber between demand and supply is about 15 million m^3. Because of the commercial importance of timber, India has started importing timber from other countries.

The use of wood and its requirement are tabulated below:

Table 8.1: Use of Wood

Region	Total Wood Consumption, billion m^3	Wood use (%)	
		Industrial	Fuelwood
World	3.2	46	54
Developing countries	1.8	18	82
Developed countries	1.4	84	16

Table 8.2: Requirement of Wood (in Metric ton)

Category	Requirement (1980)		Requirement (2000)		Addl. req.	
	m^3	% of total	m^3	% of total	m^3	% increase
Fuel wood	188.6	87.5	225	78	36.4	19
Timber	22.7	10.5	46.8	16	24	106
Pulp wood	4.2	2.0	17.7	6	13.5	323

Source: FAO Year Book of Forest Products (1981).

Deforestation has not only deprived us of economic benefits but brought in its trail environmental disaster—increased carbon dioxide content in the atmosphere and hence increased Green-house Effect, soil erosion, drought and flood, loss of habitats of wild birds and animals and also loss of valuable species including medicinal plants.

As forests are renewable resources, priority should be given to afforestation, i.e. plantation of new trees to make up the loss due to deforestation. The regeneration cycle of forests can vary widely in different climatic regions from 15 years to 80 years. Afforestation is intended to ease the present pressure on forest resources and sustain the environment. It is also essential to control utilization and consumption of forest wealth.

(iv) **Wild Life Resources:** The bounties of Nature include wonderful variety of flora and fauna with their habitats in different types of forests. but deforestation has destroyed many habitats in the past whereby many plants and animals are on the way to extinction.

Wild life populations have been considerably depleted because of reckless hunting of animals and birds for food, furs, feathers, skins, tuska etc. since the 1950s. Thus out of 65,000 species of animals almost 600 have been identified as endangered species. We have also lost many species of colourful birds out of a total number of 1200. This wave of man-made extinction has already upset the ecological balance in the environment and badly damaged India's economy.

(v) **Fish resources:** Fresh and marine waters provide fish, an important renewable resource. It is a well-known food item with rich protein content. Fish oil is a useful tonic and fish residues are used as fertilizers. People in developing countries cannot afford meat for their diet and so they depend on fish as the only source of animal protein in their diet. In the earlier days fish was used on a small scale but now with population explosion, it has been transformed into a large scale industry. There are abundant fish resources in the country with about 30,000 species of fish.

But in recent years fish resources as a whole have been hit by increasing water pollution. This has caused fish scarcity in general.

Over-consumption and over-fishing also have turned fish like halibut and salmon into endangered species. World-wise annual catch of fish is about 8.5 crore metric tons. The major fishing grounds are the North Coast of USA (the Grand Banks) and North Sea (Dogger Banks)—these centres conduct major trade in fish products. Japan and Norway are the principal fishing nations in the world. Japan is in the forefront accounting for 26.5 per cent of the total catch. In India fishing is done in inland waters—rivers, lakes, ponds etc. to satisfy the requirements for local consumption. Marine fishing is practised in coastal areas but is handicapped due to shortage of modern equipments like trawlers and large ships and also lack of training and facilities.

Recently the government is paying attention to the development of pisciculture or fish culture by setting up fisheries in lakes, ponds and rivers. Over-fishing control and also control of water pollution are essential for successful pisciculture.

(vi) **Agricultural Resources:** From ancient times India has been predominantly an agricultural country. Hence agriculture is the pillar of Indian economy. Agricultural resources are man-made using land, soil and water as the basic resources. Production of a variety of crops provide us with our staple food and fruits. Food crops are wheat, paddy, maize and barkey. Fruit trees yield mango, banana, coconut, orange, apple, grape etc. In addition to these there are plantation crops such as tea, coffee and rubber; fibre crops like cotton and jute and commercial crops such as tobacco, sugarcane, sugar, beet etc. Each crop needs specific type of soil, climate and environment for its cultivation.

At present agriculture in India involves about 70 per cent of people and 47 per cent of total land area of the country (cf; USA twice the size of India and less than half population engages 2.4 per cent of people and 20 per cent cropland). Our production of foodgrain has gone up from 50 Mt (1950-51) to 155 Mt in 1984-85 and upto 250 Mt in 2000 A.D. The production is about three times the soil capacity (75 Mt). This has been possible with the application of advanced technology—high-input high-output, increased use of fertilisers and pesticides and extensive irrigation resulting in *Green Revolution.* Punjab and Haryana are the leading beneficiaries of Green Revolution with record production of wheat. However, this western model of high-input high-output agriculture, inspite of paying rich dividends in terms of production and making the country self-sufficient in food production, has brought in its trail a downward trend since the last decade. The latter is due to loss of fertility of soil, loss of top soil, soil degradation, depleting groundwater reserve as a result of heavy withdrawal for irrigation, increased surface water pollution etc.

The present crisis can be overcome with the help of advanced techniques e.g. organic farming, biotechnology, less use of chemical fertilisers and pesticides etc. As a matter of policy, deforestation should be reduced, afforestation should be enhanced, population time bomb should be defused while over-consumption of food and agricultural crops should be controlled. This will help in bridging the gap between agricultural yield and its potential yield and demand for food by increasing population.

(vii) **Animal Resources:** Domesticated animals providing us with meat and milk products (animal protein) are known as animal resources. They supplement our food from agriculture. They also supply us with our requirements for wool, hides and skin.

In India animal rearing and animal breeding have not received due attention and have not been commercially developed. Animals are used to supplement human labour for agriculture. For example, cattle and buffaloes are used for ploughing, drawing water from wells and pulling carts. This is an unfortunate abuse or animal resource. India happens to have the largest cattle population in the world but due to negligence and misuse, milk yield is poor on an average and is not available to the poor sections of the population. The net result is that poor children are denied milk supply and suffer from mal-nutrition.

However, in urban centres dairy units have been set up to supply dairy milk, butter, cheese and milk powder to the urban population from their commercial centres in Gujrat and Maharashtra. Networks of Mother Dairy and Metro Dairy products have dominated the urban market.

(viii) **Water Resources:** "One can live several weeks without food, several days without water but only a few minutes without air."

Like air, water is an essential constituent of the life-support system. India is blessed with good rainfall, about 200 cm. per year. The quantum of water is about 6 trillion m^3 which is among the largest in the world for a country of comparable size. But due to wastage and inefficient water management along with unequal distribution of rainfall, a large part of the country suffers from water scarcity while during the dry months even place like Cherrapunji in Meghalaya having heaviest rainfall in the world (1200 cm) experiences water shortage. Again during rainy season surface water from rivers runs off rapidly causing floods in Assam, W. Bengal and Bihar.

Among the major consumers of water is the agricultural sector. Table 8.3 shows that irrigation (inclusive of livestock) and power use 79.6 per cent and 13.7 per cent water respectively, followed by domestic (3.6 per cent) and industrial (3.2 per cent) uses.

With increasing population the demands of water from each sector will also increase further (at least by 25 per cent by 2025 A.D.). Moreover, surface water pollution is likely to increase within the same period. These will put severe constraint on water resource in near future. At present 480 million of India's 1 billion people do not get safe drinking water. Only a sound Water Resource

Management policy can save India in the foreseeable future.

Table 8.3: Water Consumption in India (2000 A.D.) Available water is 1900 million m³ per year.

Uses	(in 100 million m³/year)		
	Input	Consumed	Unutilised
Irrigation & Livestock	869	783	86
Power	150	5	145
Industry	35	10	25
Domestic	38	8	30
Total	1092	806	286

(ix) Conventional Energy Resources

The demand on energy are increasing with progress in human civilization. The quality of life or standard of living is linked with the quantum of energy consumption. In USA per capita energy consumption is 200 million British Thermal Units, BTU (1 BTU = energy required to raise the temperature of 1 lb. of water by 1°F), 125 million BTU in UK, 50 million BTU in Japan and only 5 million BTU in India. But generally much of the energy (about 60 per cent) is wasted. Maximum wastage is observed in power plants and vehicles.

The conventional energy resources are fossil fuel (coal, petroleum and diesel), wood, natural gas, hydroelectricity and nuclear energy. The energy, as consumed by man, is: 33 per cent from petroleum and diesel, 27 per cent from coal and 5 per cent from nuclear fuels.

Coal

Coal is substantially more abundant than oil or gas, the total reservoir being 7×10^{12} metric tons which is equivalent to 5×10^{22} calories. This is 1000 times more than the total global energy consumption from all fuels. The stock of coal is likely to last several centuries.

The natural defect of coal is that it is a dirty fuel to burn. On combustion, it emits sulphur dioxide which is an offensive gas which forms sulphuric acid in air and causes acid rain in far away places. Thus it poses environmental hazards (*see* acid rain in previous chapter). Excavation of coal from mines is followed by soil subsidence (depression) which endangers the residential areas above the coal mines. Moreover, fly ash arising from combustion of coal is a nuisance as solid waste which brings about environmental problems. Also being a solid, coal is less convenient to handle than petroleum or natural gas.

In order to overcome these problems, the developed countries use less polluting forms of coal by transforming it into gaseous, liquid or low sulphur, low ash solid fuel. In a typical case, high grade ash-free coal is produced as solvent-refined coal (SRC) by suspending pulverized coal in a solvent and treating with 2 per cent of its weight of hydrogen at a pressure of 1000 pounds per sq. inch and 450°c. The product is a semi-solid, m.p. 170°C having a calorific value of 16,000 BTU per pound. This compares well with the best grade anthracite coal.

Methanol, CH_3OH

It is a convenient liquid fuel which can be produced from coal. On a commercial scale, it is produced by the reaction of carbon monoxide, CO and hydrogen, H_2 at 50 atmospheres pressure and 250°C in presence of copper-based catalyst. The reactant (CO and H_2) are obtained from coal, oxygen and steam:

$$CO + 2H_2 \longrightarrow \underset{\text{(methanol)}}{CH_3OH}$$

15 per cent methanol makes an excellent additive to gasoline which improves fuel economy and also cuts down the emission of practically all automotive pollutants.

Petroleum or Mineral Oil

The consumption of petroleum and natural gas is maximum in the developed countries and has become the status symbol of a country. USA is the largest consumer of petroleum in the world (about 80 per cent of total energy consumption in USA).

The Industrial Revolution (1780) was initially fuelled by coal but later on preference was given to oil and gas which provide cleaner fuels and easy transportation. The world reserve of petroleum is about 800 billion barrels (1 barrel = 31.5 gallons = 120 litres) which will last for less than 100 years.

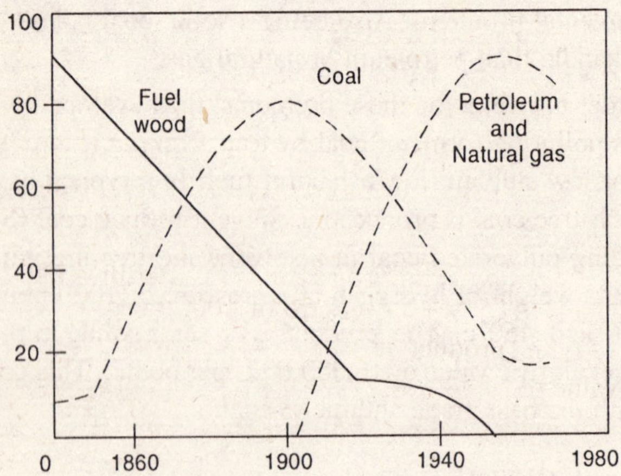

Fig. 8.1. Energy Consumption Patterns in USA

Hydroelectricity

The output from hydroelectricity (electricity from water) accounts for 21 per cent of total electricity generation, which is less than that from thermal power but greater than that from nuclear power. In Venezuela, South America 10,000 megawatts of hydroelectricity are produced which are equivalent to the production of electricity from 10 thermal power plants. In India, if water resources are properly utilised, it may be possible to generate more than 10,000 megawatts of electricity. But at present only 16 per cent or 6,500 megawatts of hydroelectricity are generated.

For generation of electricity from hydel project it is necessary to utilise energy produced from the descent of water from higher to lower level. In practice, a water reservoir is constructed by means of dam in a river for storage of water. Subsequently the stored water is released from upper level into a water-driven turbine placed at a lower level (Fig. 8.2) whereby electricity is generated. The Hydel Projects of Maithon, Pancyet and Jaldhaka are typical examples.

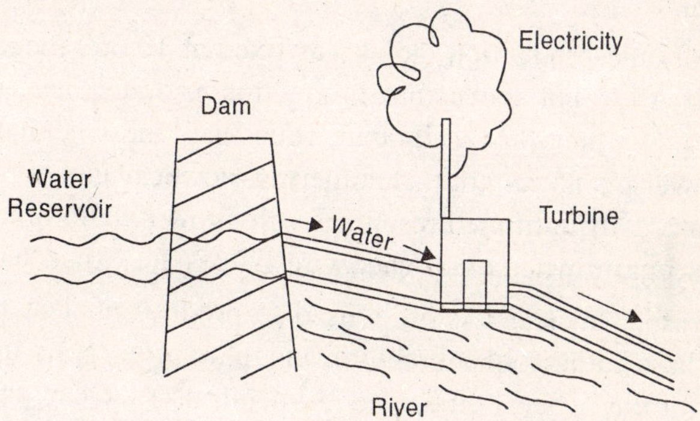

Fig. 8.2. Hydroelectricity from Hydel Project

The merits of hydroelectricity are: (1) clean source of energy; (2) no emission of greenhouse gases; (3) no consumption of fuel; (4) no need of high technology. But there are several environmental issues—flora and fauna in the region are disturbed due to construction of dam; local people become refugees as they are uprooted from their houses; the capacity of the reservoir gets reduced due to siltation; occurrence of floods in the area when surplus water has to be discharged in monsoon season. Hydroelectric dams are costly and take a long time for construction. In order to make hydroelectricity generation viable, it is necessary to adopt a long-term programme of afforestation, environmental conservation, housing, public health and transport and ensure close co-ordination among these departments.

Nuclear Power

It contributes only 5 per cent of total electricity generation. Nuclear power plants do not emit polluting gases such as carbon dioxide, sulphur dioxide, like thermal power plants. But they have some severe drawbacks, viz. they are costly and they release large quantities of radioactive fission products.

The radioactive wastes remain lethal (deadly) for thousands of years and for this no foolproof disposal method has been devised. That is why big nuclear power projects have not succeeded in the long run.

In India the production target was fixed at 10,000 megawatts by 2000 A.D., but the actual production is much less in the nuclear power stations at Tarapur, Rajasthan and Chennai. Nuclear power plants cannot match thermal power plants at present but in future, its unlimited resources will allow it to dominate the energy scenario when other energy resources are exhausted.

At present nuclear fission is used to produce nuclear power. Heavy large atoms like uranium and plutonium split up into smaller atoms when bombarded by neutrons (nuclear particles with mass 1 and zero charge). This splitting or fission liberates vast amounts of energy, which through conventional techniques is converted into electricity. Thus nuclear power is generated.

It has been calculated that 1 kg of Uranium-235 on a complete fission by slow neutrons releases energy equal to 1.7×10^{13} calories. This means energy-wise, 1 lb. of Uranium-235 ≡ 5 million lbs. of coal ≡ 20 million lbs. of T.N.T (highly explosive, chemical).

This is the secret of nuclear energy/power.

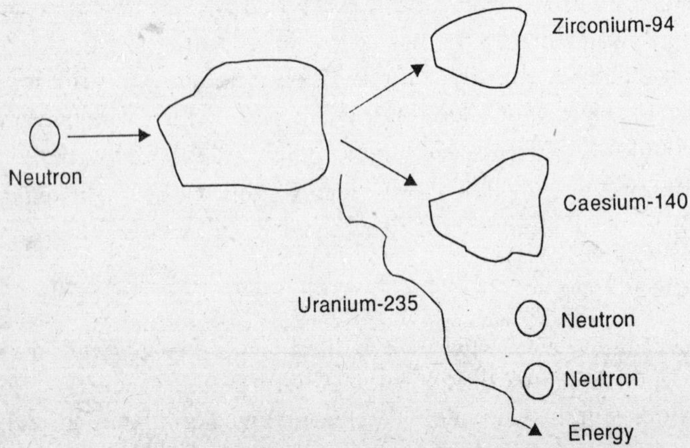

Fig. 8.3. Nuclear fission

Wood

Wood is a major renewable natural resource. The major important products are wood, paper, cellophane, rayon, plywood, plastic, particle board, turpentine, methanol, etc. In USA the production of wood and wood products is the fifth largest industry. Ideally, as in USA forests cover 38 per cent of the total land area; in India it has come down to about 15 per cent at present from 80 per cent, two thousand years ago.

It is interesting to compare between India and USA in respect of deforestation. In USA the Sunday issue of the leading newspaper, New York Times consisting of 500 pages requires 25 hectares (1 hectare = 2.5 acres = 7.5 bighas) of forest. According to an estimate, an American destroys as much forest for his needs for paper as an Indian for his domestic fuel. The value of a 50-year old tree has been estimated as about more than Rs 20 lakhs—the various functions of a 50-year old tree are roughly evaluated as follows:

(i)	Oxygen production (for 50 years)	Rs 2,50,000
(ii)	Transformation into protein	Rs 20,000
(iii)	Control of soil erosion and soil fertility	Rs 2,50,000
(iv)	Recycling of water and control of humidity and atmospheric temperature	Rs 2,50,000
(v)	Habitat for birds and other animals and insects, etc.	Rs 2,50,000
(vi)	Control of air and heat pollution	Rs 5,00,000
	Total	Rs 15,20,000

This estimate excludes the value of timber/wood for furniture, fuels, medicines, etc. which will be an extra Rs 3-4 lakhs.

Thus the tree, with its 50-year services as above, costs about Rs 20 lakhs (1980 estimate) which at present market prices, will be around Rs 40 lakhs. The public should be made aware of the value of a tree and its services to man and environment during its life time.

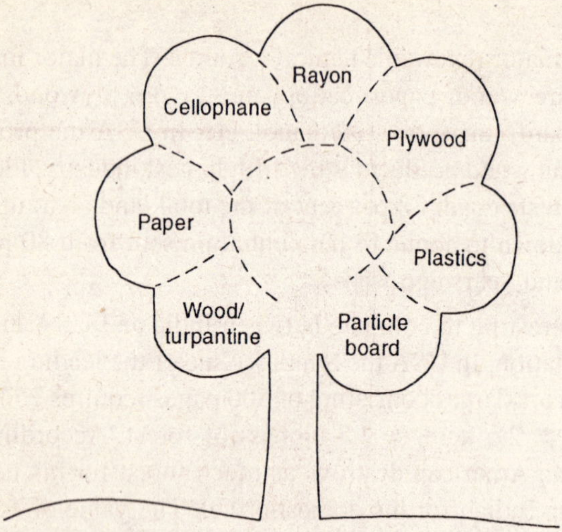

Fig. 8.4 Trees—sources of many important products

In India 76 per cent of population lives in villages—almost all of them use wood as fuel for cooking. This is the main reason for extensive deforestation in rural areas: each year we are losing about 1.3 million hectares forests. Deforestation helps increase in greenhouse gas, carbon dioxide concentration. Hence for the welfare of the country as a whole, it is essential to minimise deforestation by adopting, alternative resources of afforestation on a large scale to meet the needs of domestic fuel.

Natural Gas

It is better fossil fuel than coal and petroleum since on burning, it produces less carbon dioxide. For production of one unit of energy, mineral oil, coal and wood, on burning, produces 35 per cent, 75 per cent and 80-90 per cent more carbon dioxide than natural gas. Hence natural gas is the obvious choice as a cleaner fuel. Its reserve, however, are limited and can continue to feed only for the next 70-80 years. At present, in India the exploitable reserve of natural gas is about 700 billion cubic metres.

8.4. NON-CONVENTIONAL ENERGY RESOURCES

Solar Energy

India, being a tropical country, is blessed with abundant sunshine, 2000 kilowatt hour/sq. metre (kWh/m^2) per year for about 200-300 days in a year. The daily sunshine is between 5-7 kWh/m^2. This is an enormous and model energy resource, which is clean, pollution-free and inexpensive. It requires to be converted into other forms of energy by suitable techniques—it can meet our energy demands for ever. The solar energy, incident on earth in one week, is equivalent to the energy from the entire coal reserve of the world. Again the solar energy available on earth for 45 minutes is enough to meet our energy demand for one year.

However, the major problem is that sunlight is diffuse (wide-spread) in nature and difficult to be stored and utilized. But with advance technology, the present high costs may be cut down so that solar energy can be utilized on a large scale in future. At present, solar energy is ten times more expensive than thermal power. But with advance technology, it will be cheaper and hold the key to meet our energy demands in future.

Sunlight may be directly converted into electricity through photovoltaic cell. The latter is a device for conversion of light energy into electrical energy. The efficiency of conversion of light into electricity is only 18 per cent and it is expensive at current prices. We can use solar energy in two ways: (1) use of solar heat and (2) use of solar electricity. Use of the former permits one to boil water or dry foodgrains. Accordingly, several gadgets have been produced such as solar cooker (for cooking, solar dryer (for drying grains), solar water heater (for heating water), solar distillation (for water purification), etc. Recently there have been extensive use of these solar equipments in rural and semi-urban areas. By using the second method, i.e. solar cell, sunrays are converted into electricity. Since these solar cells are made of silicon, these are called silicon cells.

The advantages of solar photovoltaics are that they can replace systems which use diesel and they are free from chemical and noise pollution. They could be installed in remote areas in forests and deserts where installation of electric cables are cost-prohibitive.

Solar power, with government subsidy (Department of Non-conventional Energy Source, DNES, Government of India) is being used in remote rural areas in West Bengal in the forms of solar lanterns, solar streetlight and solar pump (for irrigation).

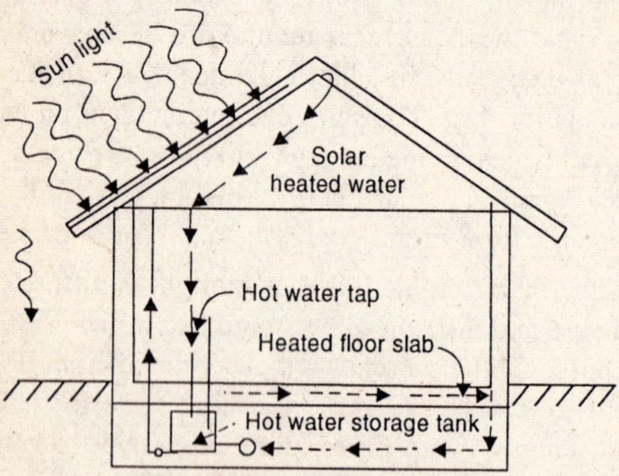

Details of heating system

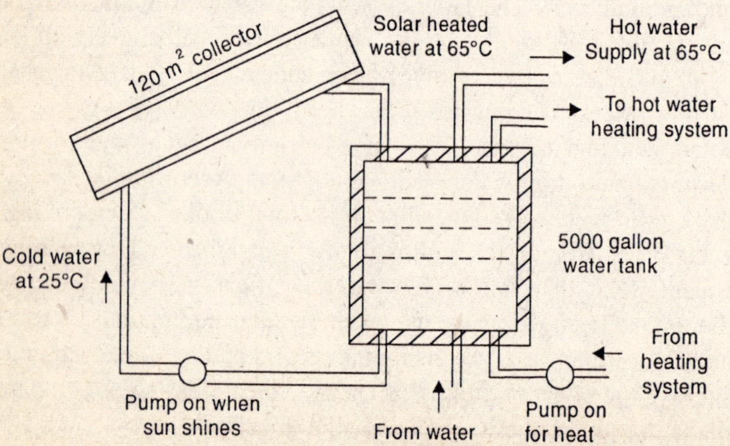

Fig. 8.5. Solar heated house

Solar powered small pumps are being used in Delhi, Haryana and Himachal Pradesh. It is desirable to use solar cookers in villages on a large scale so that extensive deforestation can be prevented. About 1 ton of wood per head per year can be saved by this process.

Figure 8.5 illustrates a detailed design for a solar heated house during winter in developed countries like USA. In these countries 20-25 per cent of fuel is consumed for providing hot water to houses and buildings. Sunlight is collected on plates in the roof and the heat transferred to a circulating water system. An average house with roof area about 1300 sq. ft. in central USA can get its energy supply for heating and hot water supply in December by this method. This may well apply to hill station houses in India in Jammu & Kashmir, Nainital, Mussorie, Darjeeling, etc. in December-January.

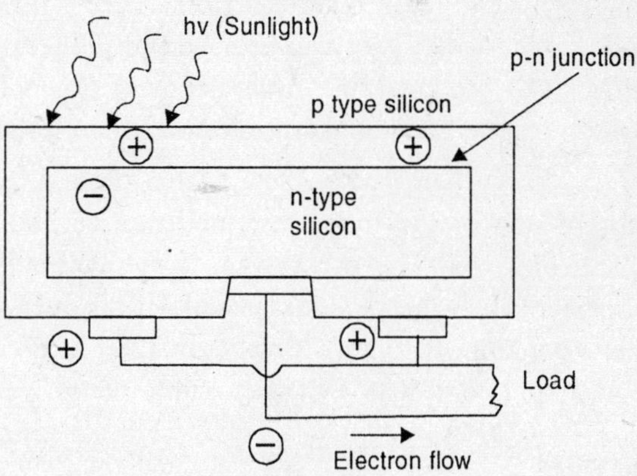

Fig. 8.6. Solar cell for electricity generation

Figure 8.6 illustrates the function of a solar cell. Light is absorbed in a plate, with the generation of positive and negative charges, which are collected at the electrodes on either side. The silicon solar cell, developed for space programmes, consists of a sandwich of a *n*-type and *p*-type silicon semi-conductors (e.g. silicon, germanium is a crystalline substance which is intermediate between a metallic conductor on one hand and non-conducting insulator on the other)—the charge separation is developed across the junction between them. *p*-type silicon conducts positive charge while *n*-type silicon conducts negative charge. The silicon cell produces electricity but is quite expensive since very high-grade crystalline silicon is required for the cell.*

Biogas

This offers an important solution to the present energy crisis in rural areas. Besides being an important domestic energy source, it offers an environmentally clean technology. There is a vast reserve of biogas in Indian villages. It is estimated that 1000 million tons of animal dung per year is available from 250 million cattle population. On an average 10 kg of wet dung is available per animal per day, which at 66 per cent collection efficiency, can yield 22,500 million cubic meters of biogas through biogas plants. This can replace kerosene oil whereby 14,000 million litres of kerosene per year can be saved in villages. Besides biogas slurries can produce 200 million tons of organic manure per year which can be good substitute for chemical fertilisers for agriculture.

The composition of the biogas is: methane, carbon dioxide, hydrogen and nitrogen. The proportion of methane and carbon dioxide varies considerable as does the calorific value. At 40 per cent methane content, the calorific value is 3200 kcal/cubic metre, while at 50 per cent, it is 4500 kcal/cubic metre.

Wind Energy

This is a cheap and clean energy resource. India, with its climatic diversity, has areas which are quite windy. According to the Indian Meteorological Department, average annual wind velocity of 6.5 metres per second at a number of places in peninsular India

---

* *n-type Semi-conductor:* Silicon lattice with arsenic as impurity contains extra electron and causes negative charge to flow.

p-type Semi-conductor: Silicon lattice with indium as impurity contains positive hole which carries current.

as also along coastlines in Gujrat, Western Ghats and parts of Central India. Such velocities are available for 6-7 months in a year.

There are some limitations for setting up of wind power mills or wind mills. They require locations where the wind velocity is at least 6.5 metres per second. In Denmark and Holland there are rows of wind mills in extensive areas and these generate 50 megawatts of electricity. A standard wind mill produces 55 kilowatts of electricity daily. Wind mills spread over extensive areas on sea shore or very high site present a beautiful scenario. Windmills prevent earthquakes where continuous wind flow causes soil erosion. In Scotland, Wales, Sweden, Germany and USA many wind mills have been constructed for cheap generation of electricity.

The technology for harnessing wind energy has become commercial in some developed countries but in India it is still in the preliminary stage. The Department of Non-Conventional Energy Sources, Government of India has installed several wind pumps with pumping capacity of 20 metres. A wind mill with a capacity to pump 400 litres of water per hour at a pumping head of 19 metres has been installed. Prospective sites are in Gujrat and Orissa on the sea coast. A 100 km stretch of coast in areas having wind speed of 10 km/hour from sea would lead to an installed capacity of 5000 megawatts. Wind energy can be used advantageously in remote rural areas and would help in saving fossil fuels.

Ocean and Tidal Energy

Ocean waves splash on ocean shores at tremendous speed—the mechanical energy in this process can be harnessed and converted into electrical energy. It has been found that in the middle of North Atlantic Ocean each wave per 1 metre height can generate 90 kW electricity whereas on the oceanshore the waves can generate 25-70 kW. During storm the generation level can rise up to 5 megawatts. Lot of research is on in this area in U.K., Canada, Norway and Japan.

In a large chamber the sea water is enclosed by oscillating water column method. Ocean/sea wave enters the chamber through an inlet pipe and forces the enclosed water upward at terrific speed—it will exert hydraulic pressure on enclosed air which in turn can rotate a turbine. Such method is expensive at present but it has immense potential which can be exploited in future with advance technology.

Tidal Wave

Tidal wave can also be tapped for generating electricity. During flow tide sea water enters river—it is possible to store such sea water in a big tank and rotate turbines by the mechanical force in the process and generate electricity. It is necessary that about 3-5 meters high sea water through flow tide enters the chamber. USSR (now CIS) and China have built small tidal power plants. In India the probable sites for exploration of tidal energy are the Gulf of Kutch and Cambay and Sundarbans and also near Andaman, Nicobar and Lakshadeep islands. the sites should be within 20-30 km from the shore in order to facilitate power transmission to the islands.

Geothermal Energy

The earth's core has a vast source of thermal energy, which has been tapped in many developed countries. In France and Hungary, hot water from hot springs has been utilised for heating houses and agricultural farms. Italy is the pioneer in this field. Later on USA, Philippines, Japan and New Zealand have been working on exploration of geothermal energy as an energy resource.

During the oil crisis period in 1973, England developed the technology for harnessing geothermal energy. If in many areas wells are dug about 5 km deep, then geothermal energy may be exploited. With advanced technology, it may be possible to generate electricity from geothermal energy in India and other developed countries.

QUESTIONS

1. What do you mean by Resources? Give examples.
2. Distinguish between Renewable and Non-Renewable Resources with examples.
3. How do you classify lands on the basis of land use.
4. Discuss the importance of top soil. Indicate annual loss of top soil in India as against world-wise loss.
5. Discuss the role of Forest Resources in Indian context.
6. Write brief note on Jhum Cultivation.
7. What are Greenhouse gases?
8. Discuss Greenhouse Effect.
9. Write notes on:
   (a) Ozone Hole
   (b) El Nino
   (c) Biogas
   (d) Tidal wave energy
   (e) Hydroelectricity
   (f) Geothermal energy
10. Comment on: "Wood is a major renewable energy resource; it offers multiple benefits to man. But it is vanishing fast."

Unit 9

Development and Environment

Over the years developmental activities have increased to meet the growing needs of ever-increasing human population. Such activities include urbanisation, housing, road construction, agriculture, industry, mining, telecommunication etc. all over the world. All these are intended to improve the quality of human life but these have been at the cost of environmental quality in general—depletion of natural resources, air pollution, water pollution, soil erosion, flood, drought, etc. which are ultimately threatening the survival of man on earth.

9.1. SUSTAINABLE DEVELOPMENT

The Policymakers of the World Conservation Strategy (1980's) introduced the term "**Sustainable Development**". However, it was not commonly used till the Bruntland Commission (World Commission on Environment and Development, headed by Norway Prime Minister, Gro Hartlem Bruntland) published its Report "Our Common Future" in 1987. According to this Report, *Sustainable Development is development that meets the needs of the present without compromising the ability of the future generations to meet their own needs.* In other words, *sustainable development means a balanced development which satisfies the* *present needs with limited utilisation and conservation of natural resources without sacrificing the quality of environment and future of the next generations.* Such balance between environment and development only lays the foundation of Sustenable Development.

Sustainable development should be part and parcel of our national plan. While noting the population explosion trend and shrinking resources base, we should practice recycling and reusing materials including food and controlling pollution so that resources are not exhausted quickly and environment does not suffer. The idea of Sustainable Development was strongly supported by the Earth Summit (1992) held at Rio de Janeiro, Brazil (United Nations Conference on Environment and Development i.e. UNCED). The Earth Summit adopted a historic Agenda 21—a comprehensive plan for leading to Sustainable Development (CED) entrusted with the task of monitoring progress in implementing Agenda 21.

The Commission has formulated the following guidelines for achieving sustainable development:

(i) Substantial reduction in population growth rates;

(ii) Housing, health care and education for the poor;

(iii) Education and empowerment of women, particularly in rural areas;

(iv) Afforestation (making up deforestation) in cities and rural areas;

(v) Use of non-conventional energy resources by industries and proper treatment of urban wastes and industrial effluents;

(vi) Conservation of land, soil, water, forest, wild life, fish and non-renewable resources by application of modern tools of science and technology;

(vii) Efficient use of resources avoiding wasteful and excessive utilisation;

(viii) A global and national strategy to eradicate poverty and

(ix) People's participation in conservation and improvement of environmental quality.

The Rio Summit through its Agenda 21 gave priority to population control, poverty eradication and people's participation and involvement in the management of natural resources.

9.2. ENVIRONMENTAL MOVEMENTS IN INDIA

In India the conservation of environment has a long history. India has a long tradition of love, respect and reverence for Nature, plants and animals. The Rigveda, Upanishads, the ancient recorded scriptures in India urge every person to conserve the five elements of the world—land, water, energy, air and space and hold respect for Mother Earth and all its living beings including plants and animals.

But we failed to learn from our history of environmental ethics as we were overwhelmed by western life style. Our concern for environment started only after more than a decade after the west took note of it and became active in handling the environmental issues. Some typical environmental movements on record are described in this section.

(a) **Bishnoi Tradition:** In Rajasthan desert area there is a small village, Khedaji where by the end of 15th century a Rajput saint founded the Bishnoi cult based on the conservation of environment. The tribal community living in this village are familiar as the Bishnois who believe in non-violence and respect for all life forms. According to them, cutting trees or killing animals is a sin.

In 1730 the Maharaja of Jodhpur ordered his men to cut some khejri trees in Khedaji village to produce lime mortar for his palace. The Bishnois adopted a new strategy at the cost of their lives. When the King's men reached the site, about 350 villagers, mostly women and children and led by a woman, hugged the trees in the forest and sacrificed their lives as they were hacked to death by king's axemen. Ultimately the king had to bow before the people's wishes. He announced prohibition of cutting trees and killing animals all over the state in the Bisnoi villages. The supreme sacrifice of the Bishnois has become a legend in the annals of forest conservation. This was the forerunner of the Chipko movement more than 300 years later. The Bishnoi tradition continues—the Bishnois have been protecting the environment of their villages all these years.

(b) **Chipko Movement:** This is the most famous and powerful people's movement in Garhwal region (Himalayan) of Uttar Pradesh. It originally started as the tribal women's protest against felling of trees by contractors but gained momentum under the leadership of Chandi Prasad Bhatt and Sunder Lal Bahuguna. It coincided with the UN Conference of Human Environment held at Stockholm (1972) which recognised the Chipko movement as a mighty Environment Protection Movement.

The term "Chipko" literally means "hugging". Dasholi Gram Swarajya Mandal (Dasholi village Self-rule Forum) spearheaded the movement in Gopeswar, Chamoli district of U.P. The DGSM under the leadership of Bhat and Bahuguna started grassroot level movement involving hill women to protect the hill ecosystem. Reckless destruction of hill forests by timber contractors caused landslides and floods in the valley of Alakananda and Bhagirathi and was destroying the fragile hill ecosystem. A unique feature of the movement was the active participation of hill women from villages who were the worst sufferers of deforestation as they had to walk 10-15 km everyday to collect their fuel for cooking. Whenever the contractors with their axmen came for cutting trees, the women hugged the trees and protected them from the axemen. The contractors withdrew from the spot and the forest was saved. Thus the entire Himalayan Garhwal region hill forests were protected from further destruction.

In course of time the Chipko movement spread all along the hill region saving all hill forests and greenery and then moved to the South in Karnataka in 1983 where it was named "Appico" movement. Soon it gained international recognition and crossed geographical boundaries to be observed as Chipko Day at new York, USA in April, 1983. A group of school children assembled at Union Square Park hugged a big tree, followed by some adults, Environmentalists from France, Germany, Sweden, Switzerland, etc. came to visit the Chipko camps and hailed the Chipko Movement.

(c) **Silent Valley Movement:** Silent Valley occupies an area of 8950 hectares at an altitude of 3000 ft. in Palaghat district, Kerala. It is surrounded by the Nilgiri forests to the north and Attappadi forests to the east—together they comprise 40,000 hectares of pristine (i.e. primitive) forest. This tropical rain forest in the Western Ghat is a precious reservoir of genetic diversity which has not been fully exploited—here plant species and other forms of life have survived for centuries in the forest. It is this gene pool to which man has to turn in future for new materials for agriculture, for life-saving drugs, etc.

The Kerala State Government decided to construct a dam in the Silent Valley for generation of 120 MW (megawatts) of electricity in 1976 at an estimated cost of Rs. 25 crores (revised in 1984 to Rs. 51 crores). The proposed dam would store 270 million cubic feet water in a reservoir spreading over 700 hectares. In order to save the Silent Valley from destruction in the process of Government dam project, the Kerala based NGO, Kerala Sastra Sahitya Parishad (KSSP) launched the Silent Valley Movement, supported by students, teachers and people of Kerala.

Soon the apex policy making bodies NCEPC, DOEn and Switzerland-based IUCN (International Union for Conservation of Nature and Natural Resources) strongly supported the cause of Silent Valley. Finally the Prime Minister (Indira Gandhi) in 1983 accepted the recommendation of top scientists and environmentalists and declared the Silent Valley as the Biosphere Reserve by cancelling the hydel project proposal of the State Government. This is the success story of an environmental movement for protection of an important biosphere reserve.

Narmada Dam

Narmada is the largest west flowing river arising from the Amar Kantaka Plateau in Shahdol district of Madhya Pradesh and travels 1300 km draining 9.88 million hectares between the Vindhya and Satpura ranges. This vast basin—average annual flow is 41 billion cubic metres—is mostly untapped because of inter-state (Gujarat, MP) water disputes. The MP Government undertook a gigantic plan—Narmada Basin Development Programme—which involves construction of 31 large dams for Narmada and its tributaries, 450 medium-sized projects and several thousand minor structures at a cost of about Rs. 25,000 crores. The benefits were projected—several million hectares of land irrigated; water supply to thousands of industries; several thousand megawatts of power, etc.

But according to environmentalists and environment action groups massive damming of the Narmada river could be a blueprint for disaster. The basin is one of the most densely forested in India. The project would imply displacement of over 1 million people, mostly tribals, submerging of over 1000 villages and over 50,000 hectares of agricultural land and also loss of forests in the region. The damage to environment and people far outweighs the projected benefits. The environmental action groups, led by the environmentalist, Smt. Medha Patekar, organised sustained movement to stall the projects of Sardar Sarobar and Narmada Sagar dams and partly succeeded.

QUESTIONS

1. What is meant by Sustenable Development?
2. Comment on Bruntland Report.
3. Write notes on:
   (i) Rio Summit (1992)
   (ii) Agenda 21 of the Earth Summit (1992)
4. Give short accounts of:
   (a) Bishnoi Movement
   (b) Chipko Movement
   (c) Silent Valley Movement
   (d) Save Narmada Movement.

Unit **10**

Environment and North-East

North-East India consists of seven states, particularly known as Seven Sisters viz. Arunachal Pradesh, Assam, Manipur, Meghalaya, Mizoram, Nagaland and Tripura. It covers a geographical area of 255,000 sq.km. which accounts for 7.5 per cent of geographical area of India. N.E. India is a strategically important area sharing international borders with four countries—Bhutan, China, Myanmar and Bangladesh. It is linked through a narrow corridor of 22 km. along Siliguri of N. Bengal. According to 1991 census, the total population of the region is 3 crores 16 lakhs i.e. 4 per cent of India's population. Figure 1 shows the map of N.E. India.

In general, the region is characterised by heavy rainfalls, rich forest cover, valleys dissected by two major rivers—Brahmaputra and Barak—and their tributaries and low population density (10 in Arunachal Pradesh to 290 persons per sq. km. in Assam. The population ratio of N.E. India to India is 41.89 i.e. 1:2.17.

10.1. FOREST RESOURCES OF NORTH-EAST INDIA

The North-Eastern States enjoy a vigorous ecosystem which favours the growth of luxuriant forests full of unique biodiversity. The favourable factors are the tropical, sub-tropical and temperate climate, heavy rainfall, abundant sunlight and humus-rich soil.

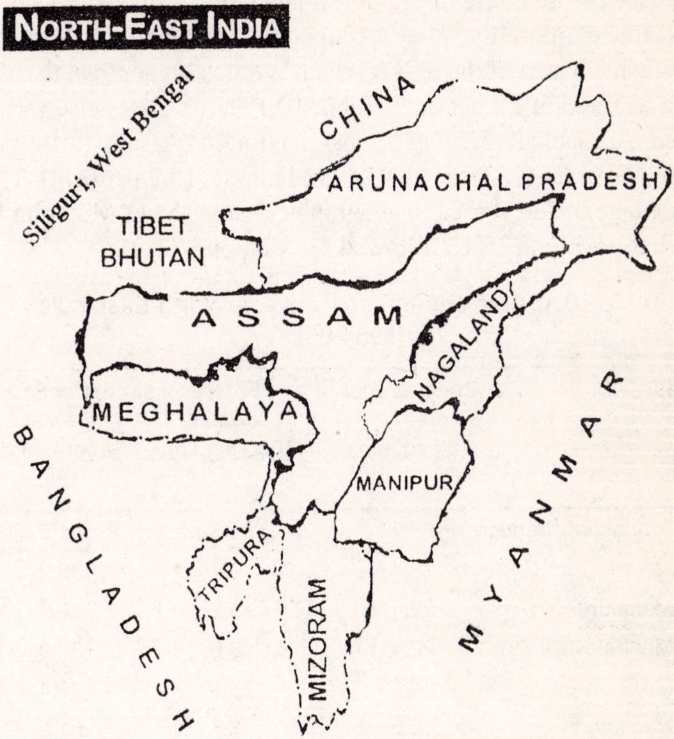

Fig. 10.1. Map of North-East India

The major forest product is forest timber which has great commercial demand for industry, railway, house building, road transport, boat and ship-building. A number of non-timber forest products—cane, bamboo, broomgrass, mushrooms, orchids, commercially important grass species, oil-yielding trees, honey, wax—are extracted from the forests each year in large quantities. Important medicinal plants such as *Texas baccata, Tinospora cordifolia, vinca rosea, Strychnos nux-vomica* etc. are also found in these forests. Gums, resins, edible wild fruits and tubers and a number of spices e.g cinamomum, lichi, large cardamom etc. contribute largely to the livelihood and economy of the people.

The forests are home to more than 200 tribal communities and hence a treasure house of biological and cultural diversity having high ethical plurality. They occupy 53.2 per cent of the total geographical area of the region which is much higher than the all-India average of 28 per cent. Table 10.1 shows state-wise forest cover: Arunachal (51.5 per cent)-maximum, Assam (30.7 per cent), Mizoram (15.9 per cent) and Manipur (15.2 per cent). The percentage of forest area to geographical area shows Mizoram in the first position (75.4), followed by Manipur (68.2).

Table 10.1: Area under Forest cover in North-East India (1984-85)

States	Geographical area in 100 sq. km.	Forest area in 1000 sq. km.	per cent of Forest area to Geographical area
1. Arunachal Pradesh	83.7	51.5	61.5
2. Assam	78.4	30.7	39.2
3. Manipur	22.4	15.2	68.2
4. Meghalaya	22.4	8.5	37.9
5. Mizoram	21.1	15.9	75.4
6. Nagaland	16.6	8.6	51.8
7. Tripura	10.5	6.3	60.0

10.2. TYPES OF FORESTS

There are eight types of forests which occur in three broadly classified zones: Tropical, Temperate and Alpian zones:

(i) Tropical evergreen forests

(ii) Tropical semi-evergreen forests

(iii) Tropical moist deciduous forests

(iv) Sub-tropical broad leaved hill forests

(v) Sub-tropical pine forests

(vi) Temperate forests and

(vii) Moist alpine forests

The forest types, their altitudinal ranges, places of occurrence along with important species are given in Table 10.2.

Table 10.2: Forest Types of North-East India

Forest types	Altitudinal ranges (km)	Places of occurrence/ Important Species
Alpine	Above 3500	Arunachal Pradesh/Rhodendron spp, Arenaria spp.
Temperate	1800-3500	Meghalaya, Nagaland, Arunachal Pradesh and Mikir hills/Pine, Oak, Willow, Birch
Sub-Tropical Pine	1000-3500	Kameng, Siang, Tirupand and Lohit districts of Arunachal Pradesh, Khasi and Jayantia hills of Meghalaya/Pinus Khasiya
Tropical evergreen	up to 900	Brahmputra valley of Assam, foothills of Eastern Himalayas/ Hollong, Hollock, Nahar Makai, Tata Sopa and Champa

10.3. FLORA AND FAUNA OF THE NORTH-EAST

Flora: There are at least 8000 plants in the region according to a recent survey. This is almost 50 per cent of the available flora within the country and most of them are endemic to this region. Among the species there are 1500 species of trees, 337 species of climbers and climbing shrubs, 700 species of fern, 800 species of monocots and 350 species of grass. While in India there are 150 species of bamboo, there are 61 species of bamboo in this region alone. Besides these, there are 2000-3000 medicinal plants, 200 varieties of citrus plants and 600 species of orchids of great horticultural value. The region is also famous for the orchid species which account for 40 per cent of the total species in the country.

Fauna: The rich flora of the region is equally matched by a wonderful variety of fauna (wild life). These include herbivores such as rhinoceros, elephant, wild buffalo, bison, antelope, deer and wild boar. The one-horned rhino of Assam accounts for huge foreign exchange earning. It is a rare species and more or less extinct elsewhere in the world. Sixteen species of monkeys are found in Assam. Among them Hoolock Gibbon deserves special mention as only three such varieties live in the world at present. Black deer, barking deer, musk deer and Mithun (Gaur) live in Arunachal Pradesh. Carnivores like tiger, leopard, black panther, bear etc. live in their ideal habitats in tropical evergreen forest and grasses. Among the reptiles are varieties of snakes such as cobra, python etc. turtles, tortoises, alligators, gharials, chameleon and lizards. Moreover, there are wonderful varieties of colourful birds and butterflies in the forests. Several wild life sanctuaries have been set up in selected sites e.g. Kaziranga in upper Assam, Manas in lower Assam, Namdapha in Arunachal Pradesh etc. Some of these, notably Kaziranga and Manas have been developed as National Parks which are important tourist centres. Some Biosphere Reserves have also been established at Manas in Assam and Nokrek in Meghalaya for conservation of genetic diversity.

10.4. DEFORESTATION

Root causes of Deforestation: (i) The high growth rate (3.0-5.0), exceeding national average of 2.72 (per 100), and density of population put pressure on forest resource base which meets food and essential requirements for people, particularly the hill tribes who lead forest-centred lives. They use firewood as fuel for cooking and heating, eat food from forest, meat from birds and animals etc. The demands of tribals as well as poorer section of villages are increasing with increase in population. These have resulted in depletion of forest wealth—high value plants, trees and animals along with the forests. In Assam immigrants have settled on forest lands in Goalpara, Barpeta, Darang and Lakhimpur districts converting these into agricultural lands.

(ii) **Jhum** cultivation is another important cause for deforestation. In North-East India about 14,600 sq.km. of forest area (6 per cent of total geographical area of the region) is under Jhum cultivation. This has caused degradation of forest land and primal habitats. Jhum practice destroys by fire much of the vegetation cover along with trees, plants and grass. According to an estimate of Indian Council Agricultural Research (ICAR), about 40 tons of top soil per hectare are eroded under Jhum cultivation, bringing about low yield of agricultural crops which cannot meet even half the needs of the farmers. But the Government cannot eradicate the practice as the land belongs to the tribals and they are unable to offer alternative means of livelihood to the Jhum farmers. Jhum practice has brought in its trail several environmental problems—loss of soil fertility, loss of burnt forests, loss of land capacity for regeneration of vegetation and low yield of crops, defeating the very purpose of Jhumming. However, the evil practice goes on for year after year.

10.5. FLOOD AND ITS CAUSES IN NORTH-EAST INDIA

Flood is the most recurring disaster in India and particularly in North-Eastern region. Among the most flood-affected countries in the world, India is second only to Bangladesh.

The Brahmputra basin is the most flood-prone region and it covers an area of 17.5 million hectares in North-East. The river along with its tributaries affects at least four states viz. Arunachal Pradesh, Assam, Meghalaya and Nagaland.

Assam is the most flood-prone state where over 75 per cent of the plains are vulnerable due to Brahmputra, Barak and their tributaries. During monsoon season (June-September) the Brahmputra spreads itself 10-16 km in width, causing devastating floods each year. Since 1986 the floods have become more and more devastating. In 1998 Assam flood took toll of 180 lives, damaged an area of 9.72 lakh hectare in 20 districts and Brahmputra remained above danger level for more than 100 days, an all-time record. The flood hit even the Kaziranga sanctuary where some animals died. Floods also dislocate transport and communication and spread water borne diseases and epidemic. In short, the entire economy of the region becomes the victim of flood.

What are the causes of recurring floods in the North-East? Natural ecological and anthropogenic factors are responsible for these floods. The Brahmputra and its tributaries originate from the Himalayas, Patkai-Naga hills, Mikir hills, Khasi and Garo hills. These hills are not only erosion-prone but also seismic (earthquake). The Brahmputra transports heavy sediment load (2.12 million metric tons per day during monsoon). Deforestation plus shifting cultivation in catchment areas of the rivers account for the heavy silt load. Other contributing factors are geological processes such as geology of the basin, topography and gradients of the tributaries and landslides, rockfalls, subsidence, overgrazing of grasslands by livestock and animals.

The government efforts for flood control include construction of dams, embankments and reclamation of swamps. In the North-East, however, these efforts lack co-ordination among the seven states. As a matter of fact, the National Committee on Environmental Planning and Co-ordination (NCEPC) pointed out that mismanagement of India's land, forest and water resources largely contribute to the environmental tragedies including flood. The North-Eastern Council, which is entrusted with the task of economic interests of the seven states, should take the initiative of co-ordinating the flood control measures of the seven states in collaboration with the Ministries of Environment in the states.

10.6. WATERSHED MANAGEMENT

Watershed generally means catchment basin i.e. an area of land that forms the drainage of a stream or river. It constitutes a big area comprising the down catchment area of the river at the foothills to the steep mountain slopes surrounding the uphill forests, which absorb rainwater and control its runoff. All watersheds are located in hill forest areas which are under the control of the forest department of the state.

Watersheds function as natural reservoir from which the rivers and their tributaries draw water. Therefore, watershed management is essential for conservation of forests, land and water. The main objective of watershed management is to store water, prevent soil erosion, regulate the flow of stream and reduce the fury of floods on the river valleys.

Thus it involves an integrated ecosystem unit that includes terrestrial and aquatic ecosystem plus man and his activities. The socio-economic aspects of an area can be improved by taking a holistic approach in respect of afforestation, enhancement and development of water resources and increase in crop and animal production. In the hilly areas the maintenance of dense forests alone can ensure water supplies during dry season. Reckless deforestation on hill slopes leads to drainage of water down the slope below so that there is no scope for water shortage a top hill areas. This accounts for the fact that Cherrapunji, the wettest place on earth, remains dry and faces acute water scarcity during dry season.

Efficient function of Water shed Management is likely to increase production of fuelwood and fodder, boost horticulture, animal husbandry and pisciculture thereby opening up avenues for job. In addition, it will also reduce downward runoff and flood disaster.

The North-Eastern Regional Institute of Water and Land Management imparts training in Watershed Management. It teaches hill farmers about conservation of basic resources—water, soil and plants resources. The programmes at the same time motivate the hill farmers to adopt an integrated approach covering crops, agro-forestry, animal husbandry, spice cultivation, etc.

10.7. WETLAND CONSERVATION

The Directory of Asian Wetlands has listed Deepor Beel as one of the 950 important international wetlands on the continent. The bird sanctuary is about 16 km southwest of Guwahati, Assam's premier city and en route to the airport nearby. Over the years, a railway line and numerous private and government buildings including small industrial units have intruded into the area, disturbing the birds and causing imbalances in the ecosystem. The wetland is also an area which functions as a natural reservoir for Guwahati, taking in lot of backflow from the Brahmaputra during monsoon.

Recently the Assam Government yielded to the demand of environmentalists and media to withdraw their order excluding Deepor Beel from protected status. This is good news for bird guests and bird lovers as the Deepor Beel has been recognised for a long time as a major wetland and nesting area for local and migratory birds.

QUESTIONS

1. Point out the strategic importance of the North East India. Illustrate with a map.
2. Give an account of the Forest Resources of North East India. List the important forest products.
3. Classify the forest types of N.E. India indicating the altitudinal ranges and places of occurrence.
4. Briefly describe the flora and fauna of N.E. India.
5. Name the factors contributing to deforestation.
6. Comment on Jhum cultivation and its harmful effects on environment.
7. Account for the recurring floods in N.E. India.
8. Write notes on:
   (a) Watershed Management
   (b) Wetland Management
   (c) Wild life resources of N.E. India
9. What are the environmental issues of N.E. India? Suggest a list of the remedial measures.

REFERENCES (UNIT 1-10)

1. **A.K. De** – Environmental Chemistry, 4th edition, New Age International Ltd., New Delhi (2000)
2. **A.K. De** – Environmental Studies, Rural Technology Centre, University of Burdwan, Chap. 1-4 (1999)
3. **A.K. De** – Environmental Studies, New Age International Ltd., New Delhi (2001)

4. **T.M. Das** – Man and Endangered Species, Shaibya Prakashan, Calcutta (1995)
5. **T.M. Das** – Is the Earth for Man Only? Jonaki Publishers, 4th ed., Calcutta (1995)
6. **T.M. Khosoo** – Environmental Priorities in India and Sustenable Development, Indian Science Congress Association, New Delhi (1986)
7. **D.D. Monte** – Temples or Tombs? Centre for Science and Environment, New Delhi (1985)
8. A.K. Agarwal (Editor) – The State of India's Environment (1984-85) New Delhi (1985)
9. **A.K. Mitra** – Urban Environment in Crisis, New Age International Ltd. New Delhi (2000)
10. **A.C. Mohapatra, S.K. Barik and C.S. Rao** – Man and Environment, Star Publishing House, Shillong (2000) 2nd ed.
11. **S. Bhattacharya** – Man and Environment, Alpha Plus Publishing House, Shillong (2000).

Unit 11

Feed Back Exercise (Objective Test)

This Feed back Exercise is intended to help the students to test their understanding of the subject matter of this book which in turn will sustain their interest in environmental studies/science. Put a tick mark ( ) within the bracket against the correct answer (s). If there are more than one correct answers, both should be marked. The answers are given at the end. (Unit wise).

UNIT 1

1. Environment means
   (a) a beautiful landscape ( )
   (b) industrial production ( )
   (c) Sum total of all conditions that affect the life and development of all organisms on earth ( )
   (d) pollution of air and water ( )
2. The elements of environment are
   (a) nitrogen, oxygen and carbon dioxide ( )
   (b) nitrogen, oxygen and sulphur dioxide ( )
   (c) oxygen, ozone and carbon dioxide ( )
   (d) atmosphere, hydrosphere and lithosphere ( )
3. Environment has two major components (NEHU, 2000)
   (a) plants and animals ( )

(b) man and nature ( )
(c) biotic and abiotic ( )
(d) physical and chemical ( )

4. Abiotic environment means
(a) living environment ( )
(b) Non-living environment ( )
(c) both ( )
(d) none of the above ( )

5. The term ecosystem was coined by (NEHU, 2000)
(a) E.P. Odum ( )
(b) E.F. Blackman ( )
(c) G.E. Hutchinson ( )
(d) A.G. Tansley ( )

6. Lthosphere denotes (NEHU, 2000)
(a) air ( )
(b) water ( )
(c) rocks and soil ( )
(d) organisms ( )

7. Hydrosphere denotes
(a) water ( )
(b) microorganisms ( )
(c) plants ( )
(d) soil ( )

8. A zone consisting of land, water and air which supports life is called (NEHU, 2001)
(a) biosphere ( )
(b) atmosphere ( )
(c) lithosphere ( )
(d) hydrosphere ( )

9. Bacteria and fungi are (NEHU, 2001)
(a) decomposer ( )
(b) producers ( )
(c) consumers ( )
(d) none of the above ( )

10. The three major living components of an ecosystem are

(a) Producers, phototrophs and consumers ( )
(b) producers, consumers and decomposers ( )
(c) producers, phototrophs and decomposers ( )
(d) consumers, heterotrophs and decomposers ( )

11. The most important organisms of an ecosystem are
(a) herbivores ( )
(b) carnivores ( )
(c) green plants ( )
(d) protozoa ( )

12. The energy flow through the ecosystem is (NEHU, 2000)
(a) cyclic ( )
(b) linear and one-way ( )
(c) both cyclic and linear ( )
(d) linear and two-way ( )

13. Graphic representation of biomass relationship between the producers and consumers in an ecosystem is called
(a) Trophic level ( )
(b) Ecological system ( )
(c) Ecological niche ( )
(d) Ecological pyramid ( )

14. Habitat (NEHU, 2000)
(a) tells us about the function of the organisms in the ecosystem ( )
(b) tells us about the place where the organisms are found ( )
(c) tells us about the size of the organisms ( )
(d) tells us about life span of the organisms ( )

ANSWERS TO QUESTIONS (UNIT 1)

1. (c) 2. (d) 3. (c) 4. (b) 5. (d) 6. (c)
7. (a) 8. (a) 9. (a) 10. (b) 11. (c) 12. (b)
13. (d) 14. (b)

UNIT 2

15. A food chain is
    (a) a group of organisms which eat the same type of food ( )
    (b) animals eating the plants ( )
    (c) series of plants and animals which are interrelated in the process of eating and being eaten up. ( )
    (d) none of the above ( )
16. The base of a food chain is always formed by (NEHU, 2000)
    (a) carnivores ( )
    (b) herbivores ( )
    (c) autotrophs ( )
    (d) primary consumers ( )
17. A food chain always starts with
    (a) photosynthesis ( )
    (b) respiration ( )
    (c) nitrogen fixation ( )
    (d) decay ( )
18. Community is (NEHU, 2000)
    (a) a group of people living in a area ( )
    (b) people belonging to a particular section ( )
    (c) organisms living in a particular type of habitat ( )
    (d) mode of interacting populations developed over a period of time. ( )
19. Energy pyramids are (NEHU, 2000)
    (a) always inverted ( )
    (b) sometimes inverted ( )
    (c) always upright ( )
    (d) all of the above ( )
20. A food web
    (a) increases variety of food at each trophic level ( )
    (b) delicately balances the interrelations amongst organisms ( )

(c) decreases variety of food but increases the quantity of food at each trophic level ( )

(d) increases variety as well as quantity of food at each trophic level. ( )

21. A group of interconnected food chain is called (NEHU, 2001)

(a) complex of food chain ( )

(b) food cycle ( )

(c) food web ( )

(d) food relationship ( )

22. Which among the following receives the maximum energy.

(a) producers ( )

(b) decomposers ( )

(c) primary consumers ( )

(d) secondary consumers ( )

ANSWERS TO QUESTIONS (UNIT 2)

15. (c) 16. (c) 17. (a) 18. (d) 19. (c) 20. (d)
21. (c) 22. (a)

UNIT 3

23. The current world population is (NEHU, 2001)

(a) four billion ( )

(b) five billion ( )

(c) six billion ( )

(d) eight billion ( )

24. Malthusian doctrine implies (NEHU, 2000)

(a) World population will go on increasing without a limit. ( )

(b) World population will slow down due to ( )

(c) World population will rise more than the rise in the world food production ( )

(d) World population will decrease due to preventive checks ( )

25. According to the Malthusian theory, population increases in
    (a) an arithmetic progression ( )
    (b) a harmonic progression ( )
    (c) a geometric progression ( )
    (d) none of the above ( )
26. The average number of persons living per square kilometer area is known as ....
    (a) community ( )
    (b) density ( )
    (c) growth rate ( )
    (d) explosion ( )
27. It is estimated that the world population may stabilise around 2050 at (NEHU, 2000)
    (a) 10 billion ( )
    (b) 12 billion ( )
    (c) 14 billion ( )
    (d) 20 billion ( )

ANSWERS TO QUESTIONS (UNIT 3)

23. (c) 24. (c) 25. (c) 26. (b) 27. (c)

UNIT 4

28. The decadal population growth of India during 1981 and 1991 was
    (a) 20 per cent ( )
    (b) 24 per cent ( )
    (c) 25 per cent ( )
    (d) 30 per cent ( )
29. The ratio of urban: total population in India in 1991 was
    (a) 20 per cent ( )
    (b) 22 per cent ( )
    (c) 25 per cent ( )
    (d) 30 per cent ( )
30. The largest metropolitan city of India (1991) was
    (a) Greater Bombay ( )

(b) New Delhi ( )
(c) Kolkata ( )
(d) Chennai ( )

31. The number of 1-million-plus cities in India in 1991 was
(a) 20 ( )
(b) 23 ( )
(c) 25 ( )
(d) 28 ( )

32. India's smallest state by population size is
(a) Arunachal Pradesh ( )
(b) Sikkim ( )
(c) Nagaland ( )
(d) Assam ( )

33. The state with the highest population growth between 1981 and 1991 is
(a) U.P. ( )
(b) Bihar ( )
(c) Assam ( )
(d) Nagaland ( )

34. At the present growth rate, by 2010 India's population will reach
(a) 1000 million ( )
(b) 1100 million ( )
(c) 1200 million ( )
(d) 1500 million ( )

35. High fertility in India is associated with
(a) early marriage ( )
(b) low literacy level ( )
(c) mother's malnutrition ( )
(d) high infant mortality ( )

ANSWERS TO QUESTIONS (UNIT 4)

28. (b) 29. (c) 30. (a) 31. (b) 32. (b) 33. (d)
34. (c) 35. (a)

UNIT 5

36. Greenhouse gases are:
    (a) CO and NO ( )
    (b) CO_2, CH_4, CFC, O_3 and H_2O vapour ( )
    (c) SO_2, SO_3 and NO_2 ( )
    (d) O_3, O_2 and H_2O_2 ( )
37. The global temperature has risen since Industrial Revolution (1860) by:
    (a) 5°C ( )
    (b) 10°C ( )
    (c) 1°C ( )
    (d) 7°C ( )
38. Atmospheric CO_2 level has increased from
    (a) 290 ppm to 330 ppm ( )
    (b) 250 ppm to 300 ppm ( )
    (c) 300 ppm to 350 ppm ( )
    (d) 330 ppm to 350 ppm ( )
39. CO_2 content of the atmosphere is likely to double by the year
    (a) 2001 ( )
    (b) 2002 ( )
    (c) 2020 ( )
    (d) 2050 ( )
40. One of the major causes of formation of ozone hole is because of the chemicals called
    (a) oxides of nitrogen ( )
    (b) chlorofluorocarbons ( )
    (c) oxides of sulphur ( )
    (d) carbon dioxide ( )
41. Ozone layer occurs in
    (a) troposphere ( )
    (b) thermosphere ( )
    (c) stratosphere ( )
    (d) ionosphere ( )
42. Average thickness of ozone layer in stratosphere is

(a) 230 Dobson units ( )
(b) 50 Dobson units ( )
(c) 1000 Dobson units ( )
(d) 100 Dobson units ( )

43. Deforestation results in
    (a) less rainfall ( )
    (b) enhanced soil erosion ( )
    (c) increased cyclones ( )
    (d) increased landslides ( )
44. The ultra-violet rays are letal because of interaction of
    (a) carbohydrates, fats and vitamins ( )
    (b) minerals, rain and air ( )
    (c) proteins, nucleic acid and pigments ( )
    (d) water, carbon dioxide and oxygen ( )
45. Desertification is caused by
    (a) tourists ( )
    (b) extensive irrigation and agriculture ( )
    (c) overgrazing ( )
    (d) development activities ( )
46. Afforestation is necessary for
    (a) soil conservation ( )
    (b) soil erosion ( )
    (c) weed control ( )
    (d) low humidity ( )
47. In the hills forest cover is affected by
    (a) overgrazing by cattle ( )
    (b) soil erosion by rain ( )
    (c) leaching of soil ( )
    (d) soil pollution ( )
48. Forest brings about
    (a) air pollution control ( )
    (b) prevention of soil erosion ( )
    (c) balance of ecosystem ( )
    (d) all the above ( )

49. Burning of fossil fuel drastically affects one of the following natural cycles:
    (a) nitrogen cycle ( )
    (b) phosphorus cycle ( )
    (c) carbon cycle ( )
    (d) water cycle ( )

50. Name the region where soil erosion from wind is maximum:
    (a) Rajasthan (India) ( )
    (b) Baluchistan (Pakistan ( )
    (c) Sahel (W. Africa) ( )
    (d) California (USA) ( )

51. Which fuel triggered off the Industrial Revolution?
    (a) oil ( )
    (b) wood ( )
    (c) coal ( )
    (d) biomass ( )

52. Where is the world's largest rainforest located?
    (a) Amazon basin ( )
    (b) Congo basin ( )
    (c) Indonesia ( )
    (d) Philippines ( )

ANSWERS TO QUESTIONS (UNIT 5)

36. (b)	37. (c)	38. (d)	39. (d)	40. (b)	41. (c)
42. (a)	43. (b)	44. (c)	45. (c)	46. (a)	47. (b)
48. (d)	49. (c)	50. (c)	51. (c)	52. (a)	

UNIT 6

53. Which gas is beneficial yet harmful to life on earth?
    (a) oxygen ( )
    (b) ozone ( )
    (c) carbon dioxide ( )
    (d) methane ( )

54. The gas essential for food synthesis by plants is:
    (a) carbon dioxide ( )

(b) oxygen ( )
(c) nitrogen ( )
(d) none of the above ( )

55. Supersonic jets cause air pollution by the thinning of
(a) carbon dioxide layer ( )
(b) sulphur dioxide layer ( )
(c) oxygen layer ( )
(d) ozone layer ( )

56. Which of the following car exhausts can lead to mental disorder?
(a) nitrogen dioxide ( )
(b) sulphur dioxide ( )
(c) lead ( )
(d) mercury ( )

57. Taj Mahal is threatened due to
(a) chlorine ( )
(b) sulphur dioxide ( )
(c) oxygen ( )
(d) hydrogen ( )

58. Absence of carbon dioxide in the atmosphere implies
(a) sub-zero temperature and absence of life ( )
(b) very hot conditions ( )
(c) life supporting conditions ( )
(d) none of the above ( )

59. Carbon monoxide is a pollutant because it
(a) inhibits glycolysis ( )
(b) inactivates nerves ( )
(c) causes haemolysis ( )
(d) all of these ( )

60. Which of the following gases in city air originates from vehicles and affects health?
(a) carbon dioxide ( )
(b) ozone ( )
(c) carbon monoxide ( )
(d) sulphur dioxide ( )

61. Which of the following is the chief source of air pollution?
    (a) automobiles ( )
    (b) burning of coal ( )
    (c) industries ( )
    (d) all of above ( )

62. Which one is the most important source of air pollution in metropolitan cities?
    (a) pesticides ( )
    (b) automobiles ( )
    (c) industries ( )
    (d) radioactive substances ( )

63. The most polluted city in the world is
    (a) Tokyo ( )
    (b) New York ( )
    (c) Kolkata ( )
    (d) Mexico city ( )

64. Bhopal gas disaster was caused by
    (a) phosgene gas ( )
    (b) carbaryl ( )
    (c) methyl isocyanate gas ( )
    (d) sulphur dioxide ( )

65. Acid rains are caused by
    (a) ozone ( )
    (b) sulphur dioxide ( )
    (c) hydrogen chloride ( )
    (d) nitrogen oxides ( )

66. Which of the following domestic activities pollutes the maximum amount of water?
    (a) toilet rooms ( )
    (b) washing and bathing ( )
    (c) kitchen washings ( )
    (d) none of the above ( )

67. Most abundant water pollutant is
    (a) detergents ( )
    (b) pesticides ( )

(c) industrial wastes ( )
(d) ammonia ( )

68. Water bodies get polluted mainly by
(a) domestic sewage ( )
(b) industrial effluent ( )
(c) agricultural run off ( )
(d) radioactive waste ( )

69. Minamata disease was due to pollution by
(a) oil spill ( )
(b) organic waste from factories ( )
(c) mercury waste from industries ( )
(d) agricultural run off containing arsenic ( )

70. DDT is a
(a) non-degradable pollutant ( )
(b) degradable pollutant ( )
(c) thermal pollutant ( )
(d) none of the above ( )

71. DDT concentration increases along the food chain at each trophic level as it moves up and the phenomenon is
(a) biodegradation ( )
(b) biomagnification ( )
(c) bioreduction ( )
(d) none of the above ( )

72. Freon, used in refrigerators and air-conditioners, is
(a) ferrous sulphate pollutant ( )
(b) chloroflurocarbon ( )
(c) zinc compound ( )
(d) metal oxide ( )

73. Major causes for soil erosion in India are
(a) tilling of land ( )
(b) deforestation ( )
(c) overgrazing and deforestation ( )
(d) drought ( )

74. Jhum cultivation is prevalent in
(a) hilly areas ( )

(b) Madhya Pradesh ( )
(c) North East ( )
(d) Andaman and Nicobar Islands ( )

75. Chernobyl disaster, the world's worst nuclear disaster, took place on
(a) 25.4.82 ( )
(b) 25.4.86 ( )
(c) 25.4.83 ( )
(d) 25.4.90 ( )

76. Nuclear fall out contains
(a) iodine-131 ( )
(b) zinc-65 ( )
(c) strontium-90 ( )
(d) ruthenium-102 ( )

77. Noise pollution becomes hazardous above
(a) 90 decibel ( )
(b) 100 decibel ( )
(c) 120 decibel ( )
(d) 150 decibel ( )

78. Lead pollution in Indian cities is mainly due to
(a) Lead pencils ( )
(b) lead batteries ( )
(c) automobile exhausts ( )
(d) electronic gadgets ( )

79. Newspaper contains one of the following elements which are toxic:
(a) cadmium ( )
(b) lead ( )
(c) manganese ( )
(d) mercury ( )

80. Which is the major source of harmful radiations in the house?
(a) tubelight ( )
(b) colour TV ( )
(c) heater ( )
(d) micro oven ( )

ANSWERS TO QUESTIONS (UNIT 6)

53. (b) 54. (a) 55. (d) 56. (c) 57. (b) 58. (a)
59. (c) 60. (c) 61. (a) 62. (b) 63. (b) 64. (c)
65.(b), (c) and (d) 66. (a) 67. (c) 68. (a) 69. (c)
70. (a) 71. (b) 72. (b) 73. (c) 74. (c) 75. (b)
76. (c) 77. (c) 78. (b) 79. (b) 80. (b)

UNIT 7

81. Mega-diversity countries are
    (a) countries having huge trees ( )
    (b) countries having strong economy ( )
    (c) countries having high biodiversity ( )
    (d) countries having little forests ( )
82. *In-situ* conservation measure is
    (a) conserving a species in its own habitat ( )
    (b) net house conservation ( )
    (c) laboratory conservation ( )
    (d) none of the above ( )
83. *Ex-situ* conservation is
    (a) to conserve species in its own habitat ( )
    (b) same as a biosphere reserve ( )
    (c) to keep the elements alive away from their original habitat ( )
    (d) none of the above ( )
84. How many living organisms on earth are on record?
    (a) 1.7 million ( )
    (b) 16 million ( )
    (c) 30 million ( )
    (d) nil ( )
85. What percentage of world's biodiversity India contains?
    (a) 7-8 per cent ( )
    (b) 1-3 per cent ( )
    (c) 28 per cent ( )
    (d) 50-60 per cent ( )

86. College botanical garden is an example of
    (a) in-situ conservation ( )
    (b) ex-situ conservation ( )
    (c) traditional conservation ( )
    (d) none of the above ( )
87. Which of the following does not come under the 'Threatened Species category'?
    (a) rare ( )
    (b) endangered ( )
    (c) vulnerable ( )
    (d) exotic ( )
88. Which one in the list is a bioreserve?
    (a) Kaziranga National Park ( )
    (b) Balpakram sanctuary ( )
    (c) Manas sanctuary ( )
    (d) Rajaji National Park ( )
89. Which is the world's first National Park?
    (a) Manu National Park ( )
    (b) Yellowstone National Park ( )
    (c) Rwenzori National park ( )
    (d) Yosemite National Park ( )
90. Which domestic waste in the list below has good recycling potential?
    (a) organic matter ( )
    (b) paper and card ( )
    (c) glass ( )
    (d) metal ( )
91. Silent Valley has been preserved as it has
    (a) recreational value ( )
    (b) valuable timber plants ( )
    (c) natural forest ( )
    (d) rare plants and animals ( )
92. The Earth Summit at Rio de Janeiro was concerned with
    (a) soil fertility ( )
    (b) natural resources ( )

(c) conservation of environment ( )
(d) afforestation ( )

93. Animals and plants are best protected in
(a) zoos ( )
(b) national parks ( )
(c) botanical gardens ( )
(d) sanctuaries ( )

ANSWERS TO QUESTIONS (UNIT 7)

81. (c) 82. (a) 83. (c) 84. (a) 85. (a) 86. (b)
87. (d) 88. (b) 89. (b) 90. (a) 91. (d) 92. (c)
93. (c)

UNIT 8

94. Human resource means
(a) the population of a country ( )
(b) education, life span and living standard of people ( )
(c) financial ability of man ( )
(d) none of the above ( )

95. Forests are
(a) renewable resource ( )
(b) non-renewable resource ( )
(c) non-utilitarian services ( )
(d) none of the above ( )

96. Which is the biggest energy source on earth?
(a) coal ( )
(b) ocean tides ( )
(c) minerals ( )
(d) solar energy ( )

97. Biodiversity is
(a) all forest and animal resources ( )
(b) biotic resource ( )
(c) future resource ( )
(d) all the above ( )

98. When was Green Revolution started in India by introducing high-yielding varieties of wheat?
    (a) 1950 ( )
    (b) 1966 ( )
    (c) 1972 ( )
    (d) 1978 ( )
99. The largest wheat producer in the world is:
    (a) USA ( )
    (b) former USSR ( )
    (c) China ( )
    (d) India ( )
100. The largest rice producer in the world is:
    (a) China ( )
    (b) India ( )
    (c) Indonesia ( )
    (d) Japan ( )
101. Common energy source in villages is
    (a) electricity ( )
    (b) coal ( )
    (c) biogas ( )
    (d) wood and animal dung ( )
102. Soil conservation is
    (a) conversion of sterile soil into fertile soil ( )
    (b) aeration of soil ( )
    (c) erosion of soil ( )
    (d) protection against soil loss ( )
103. Soil erosion is prevented by
    (a) afforestation ( )
    (b) increased human population ( )
    (c) overgrazing ( )
    (d) jhum cultivation ( )
104. Non-conventional energy source is
    (a) coal ( )
    (b) petroleum ( )
    (c) solar radiation ( )

(d) electricity from nuclear plants ( )

105. Best source of renewable energy is
    (a) cattle ( )
    (b) petroleum ( )
    (c) coal ( )
    (d) trees ( )
106. Non-renewable energy resource is
    (a) forest ( )
    (b) coal/petroleum ( )
    (c) water ( )
    (d) wild life ( )
107. Recurrence of floods in the plains of North India is due to excessive
    (a) siltation of dams ( )
    (b) rainfall ( )
    (c) agriculture ( )
    (d) deforestation in watersheds ( )
108. Petroleum is a
    (a) synthetic product ( )
    (b) renewable resource ( )
    (c) non-renewable resource ( )
    (d) near-exhausted resource ( )
109. Water is a natural resource which is
    (a) non-degradable non-maintenable ( )
    (b) degradable maintenable ( )
    (c) renewable ( )
    (d) non-renewable ( )
110. The seeds of plants are preserved/stored in
    (a) gene banks ( )
    (b) species banks ( )
    (c) ecosystem banks ( )
    (d) nature banks ( )
111. Silent valley is a tropical rain forest located in the state of
    (a) Manipur ( )
    (b) Kerala ( )

(c) Rajasthan ( )
(d) Assam ( )

ANSWERS TO QUESTIONS (UNIT 8)

94. (b)	95. (a)	96. (d)	97. (b)	98. (b)
99. (b)	100. (a)	101. (d)	102. (d)	103. (a)
104. (c)	105. (d)	106. (b)	107. (d)	108. (c)
109. (c)	110. (a)	111. (b)		

UNIT 9

112. Sunderlal Bahuguna gave leadership to
(a) Appiko movement ( )
(b) Chipko movement ( )
(c) Bishnois movement ( )
(d) Narmada movement ( )

113. Medha Patkar spearheaded the
(a) Chipko movement ( )
(b) Appiko movement ( )
(c) Bishnoi movement ( )
(d) Narmada movement ( )

114. Which is the first National Park to be established in India?
(a) Velvader National Park ( )
(b) Periyar National Park ( )
(c) Corbett National Park ( )
(d) Bandipur National Park ( )

115. Who form the backbone of Chipko movement?
(a) forest departments ( )
(b) women ( )
(c) students ( )
(d) political leaders ( )

116. Which country has the oldest tradition of popular movements aimed at the protection of natural and man-made environment?
(a) Italy ( )
(b) Japan ( )

(c) U.K. ( )
(d) Sweden ( )

117. When did Bhopal disaster occur?
(a) 1956 ( )
(b) 1984 ( )
(c) 1986 ( )
(d) 1980 ( )

118. The first Biosphere Reserve in India was established in
(a) Nilgiri ( )
(b) Nanda Devi ( )
(c) Little Rann of Kutch ( )
(d) Sunderbans ( )

119. Chipko movement originated in
(a) Silent Valley of Kerala ( )
(b) Kangra Valley of Himachal Pradesh ( )
(c) Tehri Garhwal of U.P. ( )
(d) Pachmari of U.P. ( )

ANSWERS TO QUESTIONS (UNIT 9)

112. (b) 113. (d) 114. (c) 115. (b) 116. (b)
117. (b) 118. (a) 119. (c)

UNIT 10

120. What percentage of land area in North east in under forest cover?
(a) > 60 per cent ( )
(b) 40 per cent ( )
(c) < 40 per cent ( )

121. Per capita forest cover for the North east is
(a) 0.076 ha ( )
(b) 0.052 ha ( )
(c) 0.52 ha ( )
(d) 11.0 ha ( )

122. Which of these forest types is not found in North east?

(a) tropical evergreen forest ( )
(b) tropical semi-evergreen forest ( )
(c) sub-tropical broad-leaved forest ( )
(d) tropical dry deciduous forest ( )

123. Slash and burn are associated with
(a) terrace cultivation ( )
(b) wet rice cultivation ( )
(c) shifting cultivation ( )
(d) none of the above ( )

124. Jhum cultivation is
(a) harmful to ecosystem ( )
(b) good for ecology ( )
(c) afforestation ( )
(d) none of the above ( )

125. What is meant by biodiversity hot spot?
(a) hot spring containing biodiversity ( )
(b) warm climate region on earth ( )
(c) area with high biodiversity under threat ( )
(d) none of the above ( )

126. Watershed is the boundary, often a ridge of high ground between
(a) two channels ( )
(b) two river basins ( )
(c) two streams ( )
(d) two lakes ( )

127. Which of the following phenomenon causes forest fires?
(a) lightning ( )
(b) cyclone ( )
(c) earthquake ( )
(d) rain ( )

ANSWERS TO QUESTIONS (UNIT 10)

120. (a) 121. (c) 122. (d) 123. (c) 124. (a)
125. (c) 126. (b) 127. (a)

Model Test Paper

NEHU (2000)

Man and Environment

Full Marks: 100 Pass Marks: 30

The figures in the margin indicate full marks for the questions.

SECTION-A

Tick ✔ the correct answer: 1 × 50 = 50 Marks

1. The greatest cause of species extinction is
   (a) destruction of natural habitats
   (b) too much hunting
   (c) wild-life trade
   (d) none of the above
2. By 'resource' we mean
   (a) a source of supply
   (b) a means of meeting needs
   (c) materials available
   (d) natural substance
3. Resource conservation is essential to
   (a) protect the resources
   (b) ensure regular supply of resources
   (c) serve needs of the present generation
   (d) all of the above

4. Which of the following is an alternative technology?
   (a) use of coal
   (b) use of hydel power
   (c) use of solar energy
   (d) use of gas
5. Due to increase in human population there is
   (a) rise in consumption of energy and food
   (b) increase in resources
   (c) development of resources
   (d) conservation of resources
6. The UN Conference on Environment and Development was organised in
   (a) 1965
   (b) 1970
   (c) 1972
   (d) 1980
7. The most well-known leader of the Narmada Bachao Andolan is
   (a) B.D. Sharma
   (b) Jashbhai Patel
   (c) Medha Patkar
   (d) Vandana Shiva
8. The Narmada debate is primarily associated with
   (a) Doyang Dam Project
   (b) Damodar Valley Project
   (c) Ganga Basin Project
   (d) Sardar Sarovar Project
9. The Tehri Dam Project is located in
   (a) Kumaon Hills
   (b) Garhwal Hills
   (c) Chota Nagpur Hills
   (d) Saramati Hills
10. The most important people's movement for the protection of environment in India is

(a) Jharkhand Movement
(b) Tebhaga Movement
(c) Kisan Movement
(d) Chipko Movement

11. The world's largest river island is
(a) Subansiri
(b) Dibang
(c) Majuli
(d) Pashighat

12. Shifting cultivation has been the most common method of agricultural practice among the tribes of
(a) Chota Nagpur region
(b) Chhatisgarh region
(c) North-Eastern region
(d) Southern region

13. From which part of the country do maximum migrants come to the North-East?
(a) Rajasthan
(b) Punjab
(c) Bihar
(d) West Bengal

14. Which one of the following has the highest percentage of tribes in 1991?
(a) Tripura
(b) Manipur
(c) Assam
(d) Nagaland

15. Which one of the following is most diverse in terms of languages?
(a) Mizoram
(b) Nagaland
(c) Arunachal Pradesh
(d) Tripura

16. The term ecosystem was coined by
(a) E.P. Odum

(b) E.F. Blackman
(c) G.E. Hutchinson
(d) A.G. Tansley

17. Habitat
    (a) tells about the function of the organisms in the ecosystem
    (b) tells about the place where the organisms are found
    (c) tells about the size of the organisms
    (d) tells about the life span of the organisms

18. Environment has two major components
    (a) Plants and animals
    (b) Man and nature
    (c) Biotic and abiotic
    (d) Physical and chemical

19. Lithosphere denotes
    (a) air
    (b) water
    (c) rocks and soil
    (d) organisms

20. Which of the following has maximum biological diversity?
    (a) Agricultural fields
    (b) Fish ponds
    (c) Estuaries
    (d) Pine plantations

21. Energy flow through the ecosystem is
    (a) cyclic
    (b) linear and one-way
    (c) both cyclic and linear
    (d) linear and two-way

22. The base of a food chain is always formed by
    (a) Carnivore
    (b) Herbivores
    (c) Autotrophs
    (d) Primary consumers

23. Community is
    (a) a group of people living in an area

(b) people belonging to a particular sect
(c) organisms living in a particular type of habitat
(d) mode of interacting populations developed over a period of time

24. Energy pyramids are
(a) always inverted
(b) sometimes inverted
(c) always upright
(d) all of the above

25. Autotrophs
(a) are self-nourishing organisms
(b) cannot make their own food
(c) depend on other organisms for food
(d) derive their food from producers

26. Currently the world population is estimated at
(a) 5 billion
(b) 6 billion
(c) 7 billion
(d) 8 billion

27. It is estimated that the world population may stabilise around 2050 at
(a) 10 billion
(b) 12 billion
(c) 14 billion
(d) 20 billion

28. Malthusian doctrine implies
(a) world population will go on increasing without a limit
(b) world population will slow down due to positive cheeks
(c) world population will rise more than the rise in world food production
(d) world population will decrease due to preventive checks

29. Most of the high population growth countries of the world are located in
(a) Asia, Africa and Europe
(b) Africa and South America

(c) Asia, Africa and South America
(d) Asia, South and North America

30. The highest population density per area is found in
(a) South Asia and South-East Asia
(b) South Asia and Europe
(c) Europe and North America
(d) South Asia and Japan

31. The decadal population growth of India during 1981 to 1991 was
(a) 23 per cent
(b) 23.56 per cent
(c) 23.85 per cent
(d) 24 per cent

32. The State that observed the highest decennial growth of population between 1981 and 1991 is
(a) Arunachal Pradesh
(b) Mizoram
(c) Assam
(d) Nagaland

33. The State with the highest level of urbanisation in India is
(a) Uttar Pradesh
(b) Mizoram
(c) Nagaland
(d) Maharashtra

34. The life expectancy at birth of the Indian population is
(a) 50 years
(b) 55 years
(c) 62 years
(d) 65 years

35. In India's population the share of the scheduled tribe population (1991) is
(a) 5 per cent
(b) 7 per cent
(c) 8 per cent
(d) 10 per cent

36. Which one of the following is a greenhouse gas?
    (a) Nitrogen dioxide
    (b) Sulphur dioxide
    (c) Carbon dioxide
    (d) Sulphur trioxide
37. Global warming is the cause of
    (a) rise in sea level
    (b) natural disasters
    (c) forest fire
    (d) loss of soil fertility
38. Ozone holes are best seen over
    (a) the Equator
    (b) South Pole
    (c) Tropic of Cancer
    (d) Tropic of Capricorn
39. Deforestation causes
    (a) loss of biological diversity
    (b) siltation of water courses
    (c) disruption in wildlife corridors
    (d) all of the above
40. Desertification is most common in
    (a) Tropical region
    (b) Temperate region
    (c) Gangetic plains
    (d) None of the above
41. Maximum air pollution in India is recorded in
    (a) Hyderabad
    (b) Bangalore
    (c) Delhi
    (d) Guwahati
42. Carbon dioxide level in the air is increasing because of burning of
    (a) firewood
    (b) fossil fuels
    (c) charcoal
    (d) all of the above

43. DDT concentration in higher levels of food chain increases due to
    (a) eutrophication
    (b) water pollution
    (c) spray of herbicides
    (d) biomagnification
44. Maximum level of noise pollution emanates from
    (a) Loud-speakers
    (b) Automobiles
    (c) Crackers
    (d) Jet aircrafts
45. Soil pollution is caused by
    (a) agrochernicals
    (b) domestic effluents
    (c) automobile exhausts
    (d) cow dung
46. The basic source of energy into the biosphere is
    (a) coal
    (b) petrol
    (c) wind
    (d) sun
47. High biodiversity in an ecosystem brings in
    (a) stability
    (b) imbalance
    (c) disturbance
    (d) balance
48. The highest biodiversity on the earth is found in
    (a) temperate forest
    (b) tropical forest
    (c) alpine forest
    (d) tundra forest
49. The 'hot spots' are areas of
    (a) hot springs
    (b) warm places
    (c) hearths
    (d) high biodiversity under threat

50. Extinction of species is
    (a) loss of genetic distinctiveness
    (b) termination of an evolutionary lineage due to demographic failure
    (c) disappearance of a particular species
    (d) all the above

KEYS TO MULTIPLE CHOICES

1. (a)	2. (b)	3. (b)	4. (c)	5. (a)	6. (c)
7. (c)	8. (d)	9. (b)	10. (d)	11. (c)	12. (c)
13. (c)	14. (d)	15. (c)	16. (d)	17. (b)	18. (c)
19. (c)	20. (c)	21. (c)	22. (c)	23. (d)	24. (b)
25. (a)	26. (b)	27. (c)	28. (c)	29. (a)	30. (b)
31. (b)	32. (d)	33. (b)	34. (c)	35. (c)	36. (c)
37. (a)	38. (b)	39. (d)	40. (a)	41. (c)	42. (b)
43. (d)	44. (d)	45. (a)	46. (d)	47. (a)	48. (b)
49. (d)	50. (b).				

Appendix 1

Glossary/Key Terms and Concepts

UNIT-1

Biosphere consists of the earth's surface, hydrosphere, atmosphere, and various living species which exist 60 metres above the earth's surface and 10,000 metres below sea level.

Ecology science that deals with the study of interactions between living organisms and their physical environment.

Ecosystem any unit that includes all the organisms which function together in a given area where they interact with the physical environment.

Environment sum total of all conditions and influences that affect the development and life of all organisms on earth.

Hydrological cycle natural cycle based on evaporation of sea water, cloud formation, and rainfall—this provides supply of fresh water.

Mesosphere located at altitude 56-100 km in the atmosphere where temperature decreases with increasing altitude.

Stratosphere located at 11-56 km in the atmosphere where temperature ranges from –56°C to –2°C and ozone is the dominating species.

Troposphere bottom layer of the atmosphere, altitude range 0-11 km and temperature 15°C to –56°C. The important species are nitrogen, oxygen, carbon dioxide and water vapour.

UNIT-2

Chemosynthesis solar energy passes from green plants to animals and the process goes on from one organism to another in the ecosystem.

Community group of interacting populations of plants, animals or microbes living in a prescribed area or habitat. Different organisms grow and live in association with each other. Plant community consists of plant populations.

Energy path the green plants absorb solar radiation and transform it into potential food energy through photosynthesis. Animals derive chemical potential energy of food from plants. Such transfer of energy from one organism to another is known as energy path.

Energy flow movement of energy through an ecosystem—from the external environment through a series of organisms and back to the external environment.

Food chain transfer of food energy from the source in green plants through a series of organisms with repeated process of eating and being eaten.

Food web interlocking pattern of food chains due to multiple food sources.

Producers autotrophic organisms, mainly green plants, which are able to manufacture food from simple inorganic substances.

UNIT-3, 4

Carrying capacity maximum population size that an ecosystem can support.

Dependency ratio ratio of people over 60 years of age and under 15 to the rest of the population.

Doubling time period within which the population of a country doubles—100 years for developed countries and 25-30 years for developing countries like India.

Environmental refugees uprooted tribal people living in forests when deforestation occurs.

Fertility rate number of babies born to 1000 women of the reproductive age group—for India, the rate is 3.0 while that for the developed countries, less than 1.5.

Human development also known as Human Resource Development, meaning the quality of life, as judged by life span, literacy and standard of living.

Population density number of people per sq.km area.

Population explosion population grows at a fast rate and the population size exceeds carrying capacity.

Population stabilisation population kept within limit at controlled rate (0.5%) as in the developed countries.

UNIT-5

Deforestation permanent removal of forest and its undergrowth.

Desertification process of formation of desert-like conditions in dry, sub-humid areas due to prolonged drought and poor rainfall over several years.

Jhum cultivation trees and bushes in hill area are cut and burnt for cultivation—this is the traditional practice of Jhum cultivation in N.E. India.

UNIT-6

Biomagnification process of magnification or amplification at each trophic level in the food chain.

Chernobyl disaster worst nuclear accident at Chernobyl, CIS (formerly, USSR) in April, 1986. The reactor exploded as a result of nuclear runaway reactions killing 2000 people and damaging several generations of people along with the environment in the region.

Environmental pollution adverse effect on the environment (soil, water, air biosphere) caused by man-made activities.

Food chain transfer of food energy from one organism to another in the ecosystem, caused by intake of food though the process of eating and being eaten e.g. human food chain has the sequence: plants → animals → man.

Trophic level of food chain: green plants belong to Trophic level I (producer); herbivores (cattle, goat etc.), Trophic level II while man belongs to trophic level IV.

UNIT-7

Biodiversity variety of living organisms within species, between species and ecosystems. Conservation of biodiversity is important functioning of the ecosystems and hence for human welfare.

Bioreserve an area of terrestrial or coastal environment representing one of the world's natural or biogeographical regions. Biospheres are meant to maintain the gene pools of the living species.

Genetic resources comprise the world of plants, animals and microbes.

In-situ conservation conservation of species within their habitats e.g. National Parks, Sanctuaries etc.

Ex-situ conservation conservation outside their habitats in genetic resource centres, zoological gardens etc. e.g. seed banks.

Sanctuaries protected areas, as in National Parks, where killing or hunting birds, animals are prohibited by law.

UNIT-8

Green revolution high agricultural yield has been achieved by application of advanced technology of high input-high output i.e. use of fertilisers, pesticides, and extensive irrigation. This has brought about Green Revolution with record production of wheat in Punjab and Haryana.

Resource form of matter and energy, useful for functioning of organisms, populations and ecosystems e.g. food, air, water, soil, plants, animals etc.

UNIT-9

Bishnoi tradition Bishnoi tribals in Rajasthan have been following their cult of non-violence and love for all life forms including plants and animals since 15th century. They resisted attempts at deforestation by king's men and even sacrificed their lives.

Chipko movement most famous people's movement in Garwal region (Himalayas) of U.P. against deforestation in the hill areas. This was spearheaded by hill women under the leadership of Chandi Prasad Bhatt and Sunderlal Bahuguna.

Narmada movement construction of dams for Narmada river and its tributaries for Narmada Basin Development Programme was strongly opposed by environmental action groups led by Medha Patkar. The Project would displace about 1 million people (tribals) and imply loss of forests and agricultural lands.

Silent Valley movement the government plan of constructing a dam for power generation in the Silent Valley of Kerala was stalled by environmental action groups and Kerala people and thus an important biosphere reserve was saved.

Sustenable development balance between environment and development which satisfies the present needs without sacrificing the quality of environment for the future generations.

UNIT-10

Fauna wild life e.g. rhinoceros, elephant, bison etc.

Flora plant species i.e. trees, shrubs, bamboo, citrus plants etc.

Watershed an area of land that forms the drainage of a stream or river and is located in hill forest area.

Wetland big pool of water resulting from backflow of river water or sea water—important natural water reservoir.

Appendix 2

The Time Scale of Evolution

Time	Event
6.5-7.00 billion years ago	The Big Bang
4.50 billion years ago	The Earth was born
3.50 billion years ago	The blue-green algae (plant) was formed
3.00 billion years ago	The animal appeared
0.0025-0.0030 billion (25-30 lakh) years ago	The man appeared

Appendix **3**

The International SI Units

The modern textbooks have adopted the System International (SI) Units. However, these have not yet been accepted in toto. In this book some SI units, universally adopted, have been used while the more familiar units, e.g. A°, atm, degree Celsius, etc. are retained.

Physical quantity	Name of Unit	Symbol
Length	metre	m
Mass	kilogram	kg
Amount of substance	mole	mol
Energy	Joule	J
Temperature	Kelvin (1°C = K-273)	K

1 Calorie (thermochemical) = 4.184 J

Appendix **4**

Drinking Water Quality Standards

In this Table the permissible limits for water quality parameters, as laid down by the United States Public Health Standard (USPH) for drinking water and Indian Standards Institution (ISI), are listed for comparison. The ISI values, available for only a few parameters, are found to be much higher than those for USPH for no valid reasons. All the units, except for pH and E. Coli, are in ppm or mg/mL.

Parameters	USPH Standard	ISI Standard
pH	6.0-8.5	6.0-9.0
Dissolved Oxygen (D.O.)	4.0-6.0 (ppm)	3.0
Chemical Oxygen Demand (COD)	4.0	–
Total dissolved solids	500	–
Suspended solid	5.0	–
Chloride	250	600
Sulphate	250	1000
Cyanide	0.05	–

(Contd.)

Parameters	USPH Standard	ISI Standard
Nitrate + Nitrite	< 10	–
Fluoride	1.5	3.0
Phosphate	0.1	–
Ammonia	0.5	–
Total hardness (as $CaCO_3$)	500	–
Calcium	100	–
Magnesium	30	–
Arsenic	0.05	0.2
Cadmium	0.01	–
Chromium (VI)	0.05	0.05
Iron (filtrable)	< 0.3	–
Lead	0.05	0.01
Mercury	0.001	–
Zinc	5.5	–
Phenols	0.001	0.005
E. Coli (Coliform cells)/ 1000 ml (Bacteriological parameter)	100	–

Appendix **5**

Toxic Chemicals in the Environment

There are a number of chemicals in the environment. Some of them are toxic while the rest are non-toxic. The toxic chemicals are discharged by industries into air, water and soil. They enter our biological system through food chain and disturb the metabolic processes, leading in some cases to fatal results.

At present there are four million known chemicals in the world and we are adding to the list 30,000 new compounds every year. Among these 60,000-70,000 chemicals are commonly used. Apart from their benefit to increasing production, living standards and health, many of them are potentially toxic.

Toxic Chemicals in Air

- Acrylonitrile, Arsenic, Asbestos
- Banzene, Beryllium
- Cadmium, Chromium, Chlorinated solvents, Coke oven emissions, Chlorofluorocarbons
- Ethylene dibromide, ethylene oxide
- Lead, Mercury
- Ozone

- Polycyclic aromatic hydrocarbons
- Sulphur dioxide

Toxic Chemicals in Water

- Trace Elements: Arsenic
- Beryllium, Boron
- Cadmium, Chromium, Copper
- Fluoride
- Lead
- Manganese, Mercury, Molybdenum
- Selenium
- Zinc
- Pesticides (from agricultural run-off)

Appendix 6

The Haves and Have-nots
(Developed and Developing Countries)

80 : 20 Ratio 20 per cent of the world population (developed countries) consume 80 per cent of the world's resources (natural), generate 75 per cent of total solid wastes and are responsible for 70 per cent of global environmental damage.

USA, with about half of India's population and double of India's area, discharges 5000 million tons of carbon dioxide every year—five times more than India.

Al Gore, ex-Vice President of USA—UNEP magazine vol. 6, No. 2, 1994—The North (developed countries) accounts for overconsumption of resources while the south (developing countries), for population explosion and poverty.

Appendix 7

One Earth, One Family

(Global Village)

The UNEP (United Nations Environment Programme) theme for 1994 was One Earth, One Family:

The Earth is not merely a "hotel"—it is our "home".

It is a place, where we should not use the resources and then move to another spot for further exploration.

—it is a place that we should cherish and need to protect.

It is not merely a "resource" or "retreat", which we use, enjoy and then leave.

—it is our **Mother.**

Appendix **8**

India's Rank in the World

Parameter	Rank
Population	2 (next to China)
Area	7
Value added in Agriculture	5
Electricity generation	8
Value added in manufacturing	14
Gross Domestic Product	15
Export of Goods and Services	30
Human Development Index	134

Index